A NEW FIRST CHEMISTRY COURSE

Other publications from Stanley Thornes:

SCIENCE COMPANIONS – for Key Stage 3   A. Porter, M. Wood and T. Wood
A FIRST BIOLOGY COURSE   P. T. Bunyan
A FIRST ELECTRONICS COURSE   R. B. Arnold
A FIRST HOME ECONOMICS COURSE   V. Reynolds and G. Wallis
A FIRST PHYSICS COURSE   R. B. Arnold

A NEW FIRST CHEMISTRY COURSE

E N Ramsden BSc, PhD, DPhil

Stanley Thornes (Publishers) Ltd

Published by
Stanley Thornes (Publishers) Ltd
Old Station Drive
Leckhampton
CHELTENHAM GL53 0DN

First edition 1980
Reprinted 1980, 1981, 1983, 1984, 1985
Second edition 1987
Reprinted 1988
Reprinted 1990
Reprinted 1991
Reprinted 1992

British Library Cataloguing in Publication Data

Ramsden, E.N.
    A new first chemistry course.—New ed.
    1. Chemistry
    1. Title   II. Ramsden, E.N. First chemistry
    course
    540   QD33

    ISBN 0–85950–758–0

Typeset by Tech-Set, Gateshead, Tyne & Wear
Printed and bound in Great Britain at The Bath Press, Avon

Contents

Preface

This book is a second edition of *A First Chemistry Course*. It is intended for pupils who are new to chemistry. It can be used in the study of chemistry as a separate subject or as the chemical part of a general science course. The book will take pupils up to the beginning of an examination course. In some schools the book will cover a period of three years between the ages of 11 and 14 years. In other schools, it will form a third year bridge between a general science course and a GCSE course.

The experiments form the backbone of the book. They are presented in detailed form, with the steps which the pupil must take clearly numbered. A discussion of results is included, but, in order to maintain a basis of discovery by the pupils, the experimental sections are kept separate from the rest of the text. Pupils can therefore work through an experiment and form their own conclusions before checking their results either with their teacher or against the discussion in the text. All the experiments may be done by pupils except for a small number of demonstration experiments. A very large number of experiments use only the simplest of apparatus. The second edition includes some open-ended experiments and also some opportunities for pupils to design their own experiments and carry them out subject to their teacher's approval.

The applications of chemistry and its social and environmental aspects receive more attention in the second edition. Additional reading material on benefits of the chemical industry such as fertilisers, glass, concrete and photography are included. In the chapters on air and water, aspects of pollution receive more coverage, and the role of chemistry in combating pollution is explored.

E. N. Ramsden
Hull 1987

Acknowledgements

I would like to thank the following people who have kindly supplied photographs for inclusion in this book:

Biochemistry Department, University of Oxford (Figure 1.1)
Blue Circle Company (Figures 6.5, 6.12, 7.12 and 7.13)
British Museum (Natural History) Geological Museum (Figure 1.9)
British Oxygen Company (Figures 5.11, 5.12, 5.13, 5.14 and 5.15)
British Petroleum Ltd. (Figures 5.24 and 7.3)
Chubb (Figure 7.21)
Distillers Company (Figure 2.8)
Ferranti (Figure 4.1)
Martin and Dorothy Grace (Figure 8.16)
ICI Ltd. (Figures 2.1 and 5.18)
Professor D. A. Jones (Figure 5.16)
National Coal Board (Figure 7.25)
Oxfam (Figure 8.8)
Permutit Company (Figure 8.12)
Science Photo Library (Figure 5.10)
Shaw Abrasives (Diamond) Ltd. (Figure 7.2)
Shell (Figures 7.26, 7.28, 7.29 and 8.15)
Simson (Figure 9.1)
Dr. H. H. Sutherland (Figure 1.10)
Vauxhall Motors (Figure 5.23)
K. Waters (Figures 2.3, 3.1, 3.2, 6.1, 6.2 and 8.9)

The photographs used in Figures 7.1, 8.12 and 9.10 are Crown copyright and are reproduced with the permission of the Controller of Her Majesty's Stationery Office.

I would like to acknowledge my debt to all those who have helped me at various stages during the preparation of this book: Dr G. N. Gilmore for reading the draft of the first edition and making many valuable suggestions; Mr F. Ashcroft for his thoughtful advice on the changes in the second edition; Dr J. Bradley for Experiment 7.2; Stanley Thornes (Publishers) for the close attention which they have given to every detail involved in the production of this book. Finally, I thank my family for the help and encouragement they have given me.

E. N. RAMSDEN
Hull 1987

1 Working at Chemistry

1.1 What is Chemistry?

Over the centuries, natural curiosity has led people to find out more and more about the world. A vast store of knowledge has been built up. We call this knowledge **science**. Chemistry is the part of science that describes the **substances** of which the world is made. One of the jobs which chemists do is to work out methods of separating useful substances from the rocks in the Earth's crust. We use these substances, such as metals, oil, salt and limestone, as raw materials for the manufacture of the tools, machines, houses, clothing and other possessions which we need. Chemists also find ways of changing substances which are found in nature into new substances. These changes are called **chemical reactions**. The work involved in discovering something new is called **research**.

Figure 1.1　Young chemists at work

How do the discoveries which chemists make help us? The fertilisers which the chemical industry manufactures ensure that farmers can grow enough food for us. Chemical pesticides prevent their crops from being smothered by weeds or eaten by insects. It is the chemical treatment which the water supply receives that makes our drinking water safe. The comfortable houses we live in and the clothes we wear are made from fabrics supplied by the chemical industry. The metals used in the manufacture of the cars, boats and planes we travel in and the fuel they burn are all supplied by the chemical industry.

At some time during your life, you will probably need the help of chemistry in keeping healthy. Painkillers (e.g., aspirin), antiseptics (e.g., TCP), anaesthetics (e.g., fluothane) and antibiotics (e.g., penicillin) are there to help you. It is the use of chemicals in the treatment of diseases that makes this century the healthiest and least pain-ridden century so far.

Unfortunately, some of the processes which make the goods we need create unpleasant waste products. They contaminate some of our sources of water and air. Chemists are now tackling the job of finding ways of reducing this pollution and ensuring a healthier life for us.

Scientific discoveries have all been the result of work by trained scientists. A scientist is trained in **observation** and **deduction**, rather like a detective. As you work through the experiments in this book, you will have to make observations; that is, to notice carefully what happens in an experiment and write down what you see. You will have to make deductions, that is, to think what information you can obtain from the results of your experiments. To do experiments, you will have to work in a **laboratory** (or **lab**). There are rules for working in laboratories which all scientists observe.

1.2 Working in a Chemistry Laboratory

Here is a list of rules to be followed when working in a chemistry laboratory.

Laboratory Rules

1. Concentrate on your own experiment. Do not touch equipment which is not part of your experiment. Do not move around more than necessary. Do not run.

2. Wear safety glasses. If your hair falls forward, tie it back.

3. Follow directions carefully. Make sure you are using the right amounts of the right chemicals. If you want to do an experiment of your own, check it out with your teacher.

4. Do not taste chemicals. Smell gases cautiously.

5. When heating a chemical in a test tube, be sure you are not pointing the test tube towards yourself or another pupil.

6. Write down your observations as soon as you have made them.

7. Use clean apparatus. Wash up and tidy up after a practical lesson. Put solid waste into the bins, not into the sinks. Let hot objects cool before putting them away.

8. If you have an accident, a burn, a cut or a splash of some chemical, wash with plenty of cold water. Tell your teacher immediately.

1.3 The Bunsen Burner

The piece of apparatus we most often use for heating is the Bunsen burner. Figure 1.2 shows the design of this gas burner. It was invented in 1854 by a German chemist called Wilhelm Bunsen.

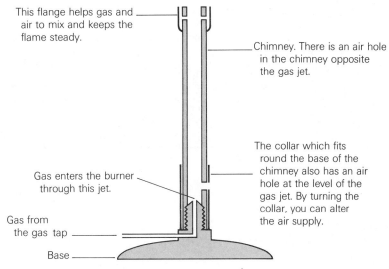

This flange helps gas and air to mix and keeps the flame steady.

Chimney. There is an air hole in the chimney opposite the gas jet.

Gas enters the burner through this jet.

The collar which fits round the base of the chimney also has an air hole at the level of the gas jet. By turning the collar, you can alter the air supply.

Gas from the gas tap

Base

Figure 1.2   The Bunsen burner

Experiment 1.1

Using a Bunsen burner

1. Connect the Bunsen to a gas tap. Close the air hole. Open the gas tap half way to the fully open position. Apply a lighted splint to the top of the chimney.

2. Note the appearance of the flame. Make a drawing.

3. Using tongs, hold a piece of pot in the flame for 1 minute. Observe the deposit which forms on the cold pot.

4. Slowly open the air hole. Notice the change in the amount of heat and light given out, the length and temperature of the flame and the noise. Draw the flame.

5. With the air hole open, repeat (3) with a fresh piece of pot.

6. Hold a piece of soft glass tubing with tongs
   (a) at the top of the flame
   (b) inside the blue cone
   (c) just at the tip of the blue cone.
   Which is the hottest region of the flame? How can you tell?

7. With the air hole open, turn the gas fully on. Note the appearance and the sound of the flame.

8. Copy and fill in this table.

Nature of flame	Air hole open	Air hole closed
Colour of flame Is it hot or very hot? Is the flame noisy? Does it flicker? Does it form soot?		

9. Compare your drawings with those in Figure 1.3.

10. Can you think up an experiment to compare the heating power of the yellow flame and the blue flame? For example, which will heat water faster, the yellow or the blue flame? Make it a fair test: everything must be kept the same except for the type of flame. When you have planned an experiment, ask your teacher to approve your plan before you try it out.

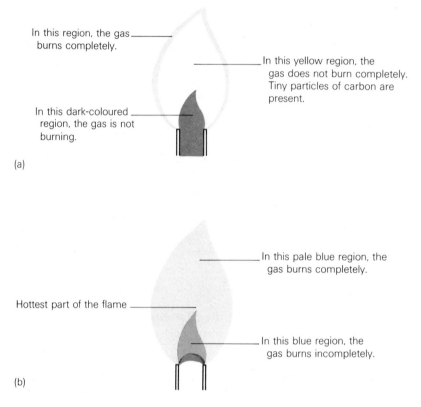

In this region, the gas burns completely.

In this yellow region, the gas does not burn completely. Tiny particles of carbon are present.

In this dark-coloured region, the gas is not burning.

(a)

In this pale blue region, the gas burns completely.

Hottest part of the flame

In this blue region, the gas burns incompletely.

(b)

Figure 1.3   The Bunsen burner flames (a) The yellow flame (b) The blue flame

Copy this Passage and Fill in the Spaces

To light a Bunsen burner, have the air hole _____ and the gas tap _____ _____. If you leave the burner alight on your bench, you want the flame to be clearly seen because _____. Use the _____ flame, with the air hole _____. To heat with the Bunsen burner, use the _____ flame. Have the air hole _____ _____ and the gas tap _____ _____. For very strong heating, use the roaring flame. Have the air hole _____ _____ and the gas _____ _____.

You can now use a Bunsen burner to heat some chemicals in Experiment 1.2.

Experiment 1.2

To find out the action of heat on some substances

1. Quarter fill an ignition tube with one of the substances listed below in step 4. Hold the ignition tube with tongs. Point it away from your neighbour and yourself.

2. Heat the ignition tube in a blue Bunsen flame. Observe carefully.

3. Write down your observations in the form of a table.

Substance heated		Observations
Name	Appearance	

4. One at a time, heat the following substances: wax, starch, ice, water, sand, salt, copper carbonate, copper sulphate crystals.

5. Look at your table of results carefully. Ask yourself which of the changes that occur on heating are easily reversed, to get back to the starting material. Which are difficult to reverse? Notice which substances do not change on heating.

1.4 The Action of Heat on some Substances; Physical and Chemical Changes

Experiment 1.2 is to find out what happens when you heat various substances. Some substances are not changed by heat. In some cases, a substance changes when it is heated, but, when it is cooled, it turns back into its original state. No new substance is formed. A change of this kind, in which no new substance is formed, is called a **physical change**.

You will have found that wax and ice simply **melt** when heated to form liquid wax and water. The change is reversed by cooling to make the liquids **solidify** or **freeze**.

We can describe these changes by writing:

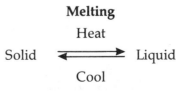

Melting

Heat

Solid ⇌ Liquid

Cool

Freezing or Solidifying

where each arrow stands for 'forms'.

Water turns into water vapour on heating. In general, a liquid turns into a vapour on heating. The process is called **vaporisation** or **evaporation**. A **vapour** is a gas at a temperature close to that at which it becomes a liquid. The change from vapour to liquid is called **condensation**.

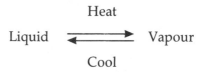

Evaporation or Vaporisation

Heat

Liquid ⇄ Vapour

Cool

Condensation or Liquefaction

All these changes are **physical changes**. The substance changes from one form into another, but no new substance is formed.

When you heat blue crystals of copper sulphate, you may notice steam coming out of the ignition tube; then a white solid is left in the tube. It looks as though a chemical reaction has occurred, to give two substances which are different from the one you started with. Let the tube cool, and add a few drops of cold water. You will see the blue colour return and feel the tube become hot. The reaction:

Heat + Blue copper sulphate $\rightleftharpoons$ Steam + White copper sulphate
            crystals   powder

will go from left to right or from right to left. It is a **chemical change** because new substances are formed. It is easily reversed.

When you heat copper carbonate, you see the green solid change to a black solid. The product looks quite different from the starting material, and in fact it is chemically different. For example, it does not fizz when put into dilute acid as does copper carbonate. You can deduce that a chemical change has taken place. This change is not easily reversed, but by a series of chemical reactions you can get back to copper carbonate.

Ask your teacher to heat some ammonium dichromate in the fume cupboard. The orange crystals change spectacularly to a solid which looks quite different. The reaction is so vigorous that you will be sure you have been watching a chemical change occurring.

These tests will have given you an idea of the difference between a physical change and a chemical change. In a **physical change**, the form of a substance changes, but it is still the same substance. In a **chemical change**, new substances are formed. The physical changes met so far are **changes of state**. Matter exists in three states, solid, liquid and gas.

1.5 States of Matter

Solid
A solid has a definite size and a definite shape. The shape may be changed when force is applied.

Liquid
A liquid has a definite size, but no definite shape. It changes its shape to fit the shape of its container.

Gas
A gas changes both its size and shape to fit its container. Gases always spread out to occupy all the space available to them. If you take a balloon of gas and 'pop' it, the gas inside it will spread out evenly all over the room. This behaviour of gases is called **diffusion**.

Figure 1.4   Solid, liquid and gas

How does matter change from one state into another? You can explain it on the basis of the **Atomic Theory**.

1.6 The Atomic Theory

Centuries ago, in 500 BC, a Greek scientist had the idea that solids, liquids and gases are all made up of enormous numbers of minute particles. He could not prove that his idea was right, and it did not catch on. A British chemist called John Dalton revived the idea in 1808. He called the particles **atoms** (from a Greek word meaning *cannot be divided*). According to Dalton's Atomic Theory, all forms of matter consist of atoms. Sometimes, a number of atoms combine to form a more complex particle called a **molecule**. Some substances are made up of single atoms; other substances are made up of molecules.

1.7 Evidence for the Atomic Theory

Dalton asked people to believe in the existence of atoms and molecules; particles which are too small to be seen, even under the most powerful microscope. It would be difficult today to find anyone who does not believe in atoms and molecules. This is because the Atomic Theory makes sense of many things which we observe in everyday life and in science.

Relationship between Solid, Liquid and Gas

The Atomic Theory helps us to understand the differences between the three states of matter: solid, liquid and gas.

In a solid, the atoms or molecules are arranged in a regular pattern. They cannot move out of position, and the solid therefore has a fixed volume and a definite shape.

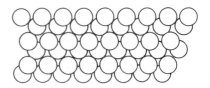

In a liquid, the molecules are in contact, but they can slide past one another. Since there is no regular arrangement of molecules, the liquid has no shape of its own. A liquid fits into the shape of its container.

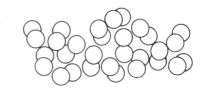

If the liquid is heated, some molecules will get enough energy to break away from the body of the liquid: they will become a gas. A gas occupies a much larger volume than the liquid which evaporated to make it. The molecules are much further apart in a gas than in a liquid.

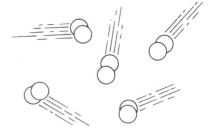

Figure 1.5   The states of matter

Dissolving

I expect you have poured boiling water on to a tea bag. You see the water gradually becoming brown as tea dissolves in it. You can imagine coloured particles splitting off from the tea leaves and moving about through the water until they are evenly spread out.

Figure 1.6   Dissolving

Dilution

If you take a bottle of orange squash and dilute it by adding water, the colour becomes paler. You can explain this if you believe that the orange squash consists of particles which spread out through the whole of the volume.

Vaporisation

A coat of paint dries in an hour or two. Where does the liquid part of the paint go? You can answer this question if you believe that the liquid consists of tiny particles (molecules). Molecules of liquid move out of the paint layer into the air around it.

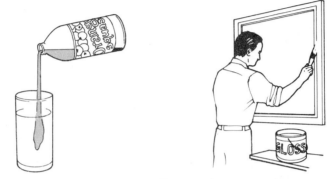

Figure 1.7   Dilution   Figure 1.8   Vaporisation

X-ray Patterns from Crystals

Many solids are crystalline. The faces of a crystal show a regular arrangement (Figure 1.9).

Figure 1.9   A crystal of garnet

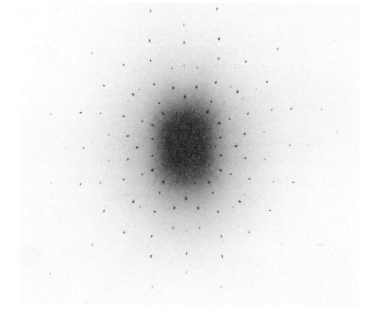

Figure 1.10   X-ray photograph of a crystal

The reason for the regular shape of a crystal is that the particles it is made of are arranged in a regular pattern. Figure 1.10 shows a pattern produced by passing a beam of X-rays through a crystal on to a photographic plate. Some X-rays pass straight through the crystal to produce a dark spot in the centre of the film. The pattern of dots is produced by X-rays which bounce off particles. Since the dots form a regular pattern, the particles in the crystal must be arranged in a regular pattern. X-ray photographs of crystals are striking evidence that solids consist of particles. Dalton would be pleased by this evidence and surprised too: X-rays had not been discovered in his day!

1.8 How Small is an Atom?

It is difficult to imagine just how small atoms are. Hydrogen atoms are the smallest. One atom of hydrogen weighs 0.000 000 000 000 000 000 000 0017 g or 1.7×10^{-24} g. There are 600 000 000 000 000 000 000 000 or 6×10^{23} atoms in one gram of hydrogen.

Every dot of ink on this page is big enough to have a million atoms of hydrogen fitted across it, side by side. If you could count all the atoms in the dot of ink, you would find that the number was more than the population of the world. A pin head measures about 1 square millimetre. Since 10 million hydrogen atoms, side by side, measure 1 mm, the area of a pin head would hold 100 million million hydrogen atoms.

1.9 The Kinetic Theory of Gases

One thing that puzzled scientists was the way in which a gas would expand or contract as the pressure on the gas was altered. To explain the compressibility of gases, they put forward the **Kinetic Theory of Gases**. The word *kinetic* is from the Greek for *moving*. The Kinetic Theory tells us that the molecules of a gas are in constant motion. The molecules travel in straight lines unless they collide with other molecules or with the walls of the container. A gas is mainly space. The actual volume of the molecules is tiny compared with the volume occupied by the gas. The molecules move closer together when the gas is compressed and move further apart when the gas expands.

Diffusion

A girl enters a room wearing perfume. How can you smell it when you are yards away from her? You can explain how on the Kinetic Theory. The perfume vaporises. Then the molecules of vapour travel through the air to reach your nose. Gases (including vapours) always spread out to occupy all the available space. This behaviour of gases is called **diffusion**.

Figure 1.11   Diffusion

Test Your Grasp of Particles

1. How can a teaspoonful of sugar sweeten a whole mug of coffee? Explain how with the help of the Atomic Theory.

2. A puddle of water slowly disappears. Where does it go? Explain with the help of the Atomic Theory. Why don't you *see* where it goes?

3. The smell of cooking wafts upstairs from the kitchen to your room. How does it reach you?

4. A balloon is filled with gas. Someone punctures it with a pin. Describe what happens to the balloon as though you could *see* what the molecules of gas were doing.

5. Why can a spoonful of food colouring be used to colour a bowlful of icing sugar?

Questions on Chapter 1

1. (a) Why is it a good idea to tie back long hair in the laboratory?

   (b) Why is it not a good idea to put bags on the bench?

   (c) Why should you stand up to do practical work?

   (d) Why is it sensible to wear goggles when you are doing experiments?

   (e) Explain why (i) liquids are not put into waste bins and (ii) solids are not put down sinks.

   (f) Why should you not point a test tube at your neighbour when doing an experiment?

   (g) What is the point of leaving a Bunsen with a yellow flame?

   (h) What is wrong with the yellow flame for heating purposes?

   (i) What is wrong with running in the laboratory?

   (j) Why is it not a good idea to eat or drink in the laboratory?

2. Write the correct word or words to fill each of the blanks in these sentences.

   (a) When water is heated, it turns into _____. This change is called _____. The reverse of this change is called _____; it can be brought about by _____. Both changes are _____ changes.

   (b) Copper sulphate crystals are _____ in colour. When they are heated, they turn _____ and give off water vapour. If water is added to the solid left behind, the colour changes from _____ to _____. The reason is _____.

   (c) Copper carbonate is _____ in colour. When it is heated, it turns _____ in colour. This change is not easily reversed because _____ substances have been formed. It is a _____ change.

3. Explain what is meant by (a) a physical change and (b) a chemical change. Give two examples of each type of change.

4. Supply the missing words in this passage.

   Solids have both a definite _____ and a definite _____. This is because the atoms in a solid are packed tightly together. A liquid has a definite _____ but no definite _____. It fits into the _____ of its container. It can do this because the molecules in a liquid are free to _____. A gas has neither a fixed _____ nor a fixed _____. The theory which describes the behaviour of gases is called the _____ Theory of Gases. According to this theory, the molecules are always _____. This is why the molecules of a gas can spread out to occupy all the space available to them. We call this behaviour _____.

Crossword on Chapter 1

Trace or photocopy this grid (teacher, please see note at the front of the book), and then fill in the answers.

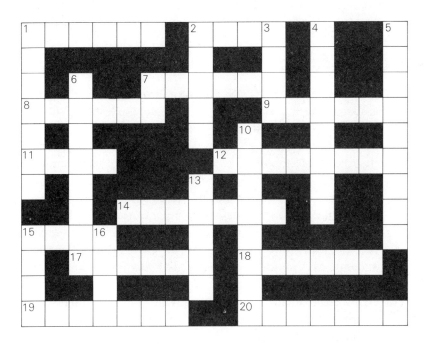

Across

1 He invented the device which we use in the laboratory for heating. (6)
2 He looks down on others (4)
7 He put forward the 8 across theory (6)
8 This theory explains changes of state (6)
9 Sit on these in the lab (6)
11 Write secret messages in this (4)
12, 18 Needed to make a new substance (8,6)
14 Spread out (especially for a gas) (7)
15, 16 down This piece of glassware is useful for lab work (4,4)
17 A continent which produces many scientists (6)
18 See 12 across
19 This theory explains the behaviour of gases (7)
20 Change from vapour to liquid (7)

Down

1 Used for weighing (7)
2 One state of matter (5)
3 Put solid waste in these (4)
4 Change from liquid into gas (8)
5 Sugar _____ to make a sweet-tasting solution (9)
6 Change from gas into liquid (8)
7 Direct current (2)
10, 18 across When this happens, no new substance is formed (8,6)
13 How frequently do gas molecules collide? (5)
15 To explain your results, _____ them over with your partner (4)
16 See 15 across

2 Methods of Separating Mixtures

2.1 Pure Substances

Timothy leapt out of bed. Slipping into his nylon shirt and polyester trousers, he ran down the acrylic-carpeted stairs, sat down at the formica-topped table, seized his stainless steel knife and fork, and attacked his egg. Apart from the egg, all these things were made by the chemical industry. To make all these chemicals, chemists use the raw materials which Nature provides. Salt, limestone, sand, coal, oil and metal ores occur naturally. From these raw materials, the chemical industry makes enormous numbers of useful chemicals. It is unusual for substances to occur **pure** in nature; most substances are found in an impure state – mixed with other substances. **A pure substance is a single substance**. Often, the first step in a manufacturing process is purification: the material which chemists want must be separated from other substances which are mixed with it. You can try a number of ways of separating substances in this chapter. They are all used to obtain pure substances from mixtures.

2.2 Filtration

Rock salt is mined in Cheshire. Figure 2.1 shows a truck laden with about 25 tonnes of salt driving through a mine. The salt is taken to the ICI factory at Winnington. There, some of the rock salt is powdered and sold for use on roads in winter. From the rest of the rock salt, pure salt is extracted. Pure salt is used in the food industry for flavouring and preserving food. It is used by ICI to make washing soda, glass and other important products.

Experiment 2.1 shows you how to separate salt from rock and sand. You tackle the problem by dissolving salt in water and then filtering. **Filtering** or **filtration** separates a solid from a liquid by allowing the liquid to pass through some porous material. Porous material acts like a fine sieve. It allows liquid to pass through but retains solid particles. You meet some new words in this experiment. When we say that a solid **dissolves** in a liquid, we mean that the solid spreads out, as tiny particles, through the liquid, so that we can no longer say which is the solid and which is the liquid. The mixture of solid and liquid is called a **solution**. The process is called **dissolving**. The solid is the **solute**, and the liquid is the **solvent**. Liquids and gases also can dissolve in liquids. A substance which will dissolve is said to be **soluble**; one which will not dissolve is **insoluble**. A solution which contains the maximum amount of solute that dissolves in it at that temperature is called a **saturated solution**. Dissolving is a physical change. No chemical reaction happens. When you evaporate the solution, you recover the solute unchanged.

Figure 2.1   A salt mine in Cheshire

Experiment 2.1

To obtain pure salt from rock salt

1. Crush and grind the rock salt in a mortar with the help of a pestle (see Figure 2.2(a)).

2. Place the solid (about 30 g) in a beaker. Add about 50 cm³ of water.

3. Stand the beaker on a tripod and gauze. Heat gently (gas tap half on, air hole half open). Stir with a glass rod to help the salt to dissolve (see Figure 2.2(b)).

4. Pour the contents of the beaker through a filter funnel fitted with a filter paper (see Figure 2.2(c)). Collect the filtrate in an evaporating dish.

5. Place the dish on a steam bath and evaporate to dryness, as in Figure 2.2(d).

6. In this experiment, as an exception to our safety rules, you may taste the product.

    Do you think that the white crystalline solid in the evaporating basin is salt?

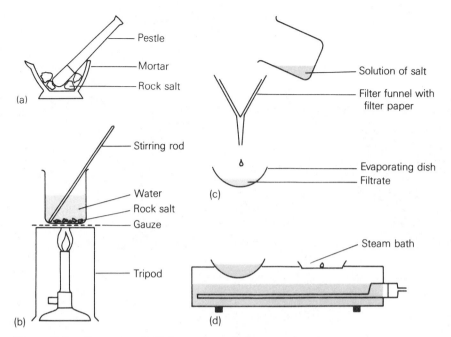

Figure 2.2   Steps in obtaining pure salt from rock salt

7. Why is the rock salt crushed in step 1?

   How do heating and stirring help in step 3?

   What is removed in step 4?

   Which steps involve (a) dissolving (b) filtration and (c) evaporation?

   How does this method of separation work? Would it work for salt and sugar? Would it work for sugar and sand?

8. How could you separate a mixture of sawdust and sand? Make a list of possible methods to try. Show the list to your teacher, and if he or she approves, try out the methods you have in mind.

Solvents other than water

Water is the most common solvent. To dissolve substances which are insoluble in water, other solvents must be used. Figure 2.3 shows some solvents.

White spirit is obtained from petroleum oil. It is a solvent for paint.

French polish is a solution of shellac in *ethanol*

The disinfectant *tincture of iodine* is a solution of iodine in *ethanol*

The rubber solution that you use to repair punctures is a solution of rubber in *trichloroethene*.

Nail varnish remover is *pentyl ethanoate*.

Trichloroethane is used in dry cleaning fluids.

Figure 2.3   Some useful solvents

2.3 Separation of Immiscible Liquids

Salad dressing contains oil and vinegar. These two liquids are **immiscible** (do not mix). They can be separated in a **separating funnel**, as shown in Figure 2.4.

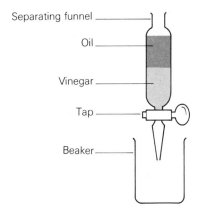

The tap is opened. Vinegar starts to run out into the beaker. When all the vinegar has gone, the tap is closed.

Figure 2.4   A separating funnel

2.4 Separation of a Solvent and a Solute from a Solution: Distillation

You can separate the solvent from a solution. You have to heat the solution so that the solvent vaporises (changes from liquid to vapour) and then cool the solvent vapour until it condenses (changes from vapour to liquid). The process of vaporisation in one part of an apparatus followed by condensation in another part of an apparatus is called **distillation**.

In Experiment 2.2, distillation is used to separate a solvent from a solute. The solvent distils; the solute remains behind.

Experiment 2.2

To find out whether ink can be separated into components

1. First try the filtration method. Does it work?

2. Try boiling some ink in a test tube. If you see that a vapour is formed, set up an apparatus in which you can trap the vapour and condense it. Figure 2.5 shows one possible apparatus.

3. Put ink into a conical flask. Drop in some pieces of broken porcelain (such as bits of a broken evaporating basin). These will help the liquid to boil smoothly.

4. Fit the flask with a well-fitting cork and a long delivery tube. This ends inside a test tube which is cooled by standing in a beaker of cold water.

5. Heat the ink with a 'half-and-half' flame.

6. Observe the distillate collected in the test tube. Does it look like ink?

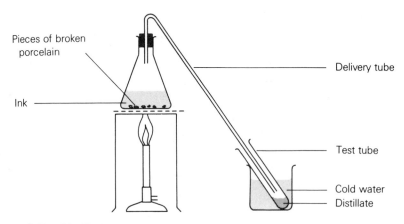

Figure 2.5   Distillation of ink

7. What is the main drawback of this apparatus? Can you suggest an improvement? Check with your teacher, and then, if he or she approves, try out your idea.

2.5 Separation of Miscible Liquids by Fractional Distillation

You will notice in Experiment 2.2 that the method of cooling the vapour is not good enough, and some steam escapes. A chemist called Justus von Liebig noticed this over a century ago. He designed a better method of cooling the vapour. It is called a **Liebig condenser** after him. Figure 2.6 shows how it works.

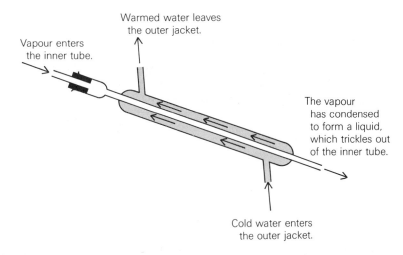

Figure 2.6   A Liebig condenser

In Experiment 2.3, **fractional distillation** is used to separate a mixture of two liquids. It is called *fractional* distillation because it separates a mixture of liquids into fractions. The distillation apparatus is shown in Figure 2.7.

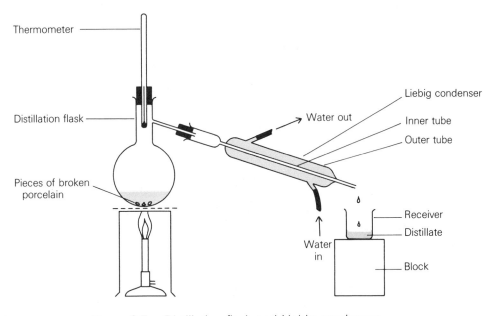

Figure 2.7   Distillation flask and Liebig condenser

Experiment 2.3

To distil a mixture of ethanol and water, using a Liebig condenser

1. Put the water–ethanol mixture into the flask shown in Figure 2.7. This is called a distillation flask because of its design. The round bottom ensures even heating of the contents, and the side-arm lets out the vapour.

2. Add some pieces of broken porcelain. Clamp the flask over a tripod and gauze. Fit a thermometer and cork, with the bulb of the thermometer just below the side-arm.

3. Bring the Liebig condenser into position carefully, and clamp it at the correct height.

4. Connect pieces of rubber tubing from the inlet of the condenser to the tap and from the outlet to the sink. Turn on the tap so that a steady trickle of water flows through the condenser.

5. Have a receiver ready to collect any distillate.

6. Heat the contents of the flask. You will see the temperature recorded by the thermometer rise until it reaches 78 °C. It remains steady at this reading while a colourless liquid distils over.

7. When the temperature rises from 78 °C, you will notice that all the ethanol has distilled over. The temperature will rise up to 100 °C, and remain at 100 °C while water distils over. Use a separate receiver to collect the water.

8. Do the following tests on each distillate:
    (a) Smell it.
    (b) Taste a drop on the end of your finger.
    (c) Take 2 cm^3 with a teat pipette, put on to a watch glass, and test with a lighted splint.

9. Copy and complete the following passage.
    Ethanol and water can be separated from a mixture of the two substances by _____. We observed four differences between ethanol and water:
    (a) _____
    (b) _____
    (c) _____
    (d) _____
    Of the two liquids, the one that vaporises more easily is _____. The reason why ethanol and water can be separated by distillation is that they have different _____.

Distillation is used in the preparation of whisky. Figure 2.8 shows the stillhouse at Ord Distillery, operated by Scottish Malt Distillers for the Distillers Company.

The method of separating two or more liquids by distillation is very important in industry. If there are a number of liquids with boiling points fairly close together, a distillation flask with a long column instead of a side-arm is used. The column provides a large surface over which evaporation and condensation can take place many times. This improves the efficiency of separating the mixture into its parts or **fractions**. Such a column is called a **fractionating column**. One of the most important users of fractional distillation is the oil industry. It separates crude petroleum oil into many fractions with different boiling points by this method (see p. 23).

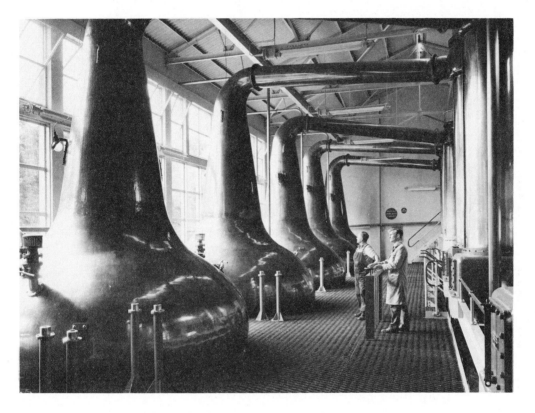

Figure 2.8   In the stillhouse at Ord Distillery, Distillers Company Ltd

Experiment 2.4

Fractional distillation of crude oil

1. Set up the apparatus shown in Figure 2.9. Push a little rocksil to the bottom of the side-arm tube. Have four ignition tubes ready in a rack.

2. Add 2 cm³ of crude oil to the rocksil. Put the thermometer into the tube as shown. The bulb of the thermometer is level with the side-arm.

3. Slant the side-arm tube slightly before you clamp it.

4. Heat the side-arm tube gently with a **small** flame.

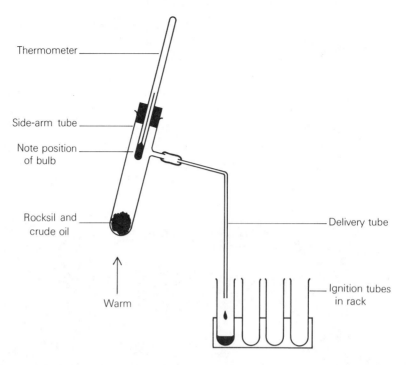

Figure 2.9   Distilling crude oil

5. Collect the fraction that distils between room temperature and 70 °C. Then change the receiver, and collect the next fraction. Collect the four fractions that distil over the ranges:
    (a) 20–70 °C
    (b) 70–120 °C
    (c) 120–170 °C
    (d) 170–220 °C

6. With each fraction, try the following tests:
    (a) Pour each fraction on to a watch glass.
    (b) Try to light each fraction with a burning splint.
    (c) Copy this table, and enter your results.

	Fraction			
	a	b	c	d
Does it run easily or is it *viscous* (syrupy)?				
Does the fraction burn quickly or slowly?				
Is the flame smoky or *luminous* (giving out light)?				

Petroleum oil is a mixture of a large number of substances. Most of them are hydrocarbons, compounds of hydrogen and carbon. Crude oil is separated into a number of different fractions by fractional distillation. Figure 2.10 shows how vaporised oil is fed into a long **fractionating column**. The compounds with low boiling points find it easier to remain gases than do the compounds with high boiling points. The low boiling point compounds pass to the top of the column, while the higher boiling point compounds condense and trickle down the column. Perforated plates all the way up the column ensure good contact between the gases going up and the liquids trickling down. The result is that fractions with different boiling points can be tapped off from different heights of the column. Each fraction is not a pure compound but a mixture of compounds with similar boiling points.

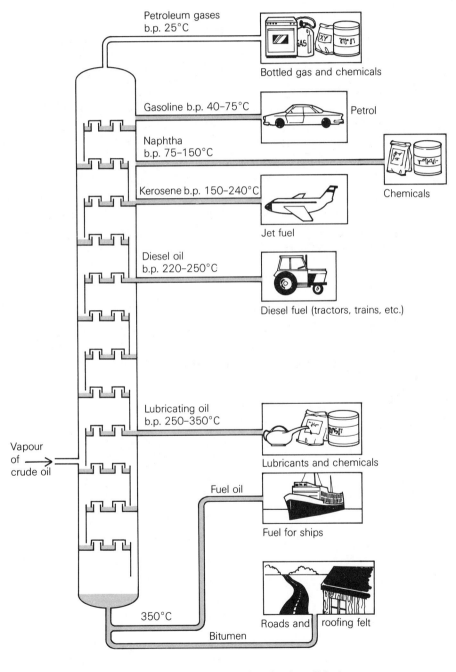

Figure 2.10   Fractional distillation in the oil industry

By means of fractional distillation, crude oil is separated into a number of very useful products. Petroleum gases are sold as bottled gases, such as Camping Gaz. Different fractions are used as fuels for motor vehicles, aeroplanes, trains and ships. Lubricating oil helps the engines to run smoothly. Wax is used to make candles and furniture polish. Bitumen is used to tar the roads.

2.6 Chromatography

In Experiment 2.2, you found that ink is not a single substance. Ink can be separated by distillation into water and a pigment. It is interesting to find out whether the pigment is a pure substance or whether it can be split up into a number of substances.

You can try the method of **chromatography**. In this method you separate substances by passing their solutions through solids such as chalk or paper. It was discovered by a Russian chemist called Tswett. He worked chiefly on the extraction of coloured pigments. These gave the process the name of chromatography since *chromos* is Greek for *colour*. The reason why the substances separate is that each substance travels through paper or chalk at its own pace. If one substance is chemically attracted to chalk or paper, this will slow it down. If one substance is more soluble than others in the solution, it will tend to move quickly with the solvent and not stick to the paper. Travelling at different speeds, like competitors in a race, the substances become spread out, one in front and others spaced out behind.

Experiment 2.5

To find out whether the pigment in black ink is a pure substance

1. Carefully put a spot of black ink at the centre of a filter paper. Use a teat pipette as shown in Figure 2.11 (a).

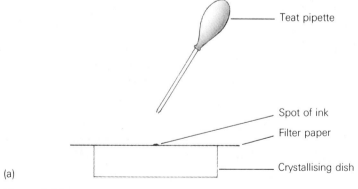

(a)

Figure 2.11

2. Place the filter paper on a glass dish and carefully add one drop of water to the spot, using a clean teat pipette as in Figure 2.11 (b).

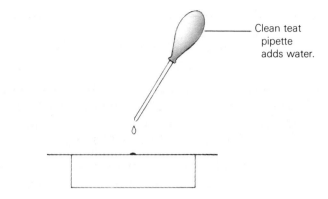

Clean teat pipette adds water.

(b)

Figure 2.11

3. Wait until the water has stopped spreading, and then add a further drop of water.

4. Repeat step 3 until you are satisfied that the change you are watching is complete. It is most important to add the water to the centre of the spot and not to add it quickly. Be patient!

5. Make a note of everything you see as you add the drops of water.

6. Dry the filter paper in the oven. Keep it to staple into your book.

Now try the method shown in Figure 2.12 (a).

Tongue
Water

(a)

Figure 2.12

1. Make two cuts from the edge of a filter paper to the centre to produce a strip 6 mm wide. Bend it down so that the strip dips into the water when placed in the dish, as shown in Figures 2.12 (a) and (b).

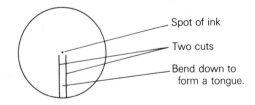

Spot of ink

Two cuts

Bend down to form a tongue.

(b)

Figure 2.12

2. Carefully put a spot of black ink at the centre of the filter paper.

3. Wait. Make a note of everything you see.

4. When you are satisfied that the change you are observing is complete, dry the filter paper. Keep it to staple into your book.

Experiment 2.6

To investigate the pigments in felt-tip pens by means of paper chromatography

1. Take a strip of chromatography paper 12 cm × 6 cm, and rule a pencil line 2 cm from the short side. Put two spots of felt-tip pen on the pencil line, 2 cm apart, and write the colour in pencil under each spot. Write your name in pencil at the bottom (see Figure 2.13 (a)).

2. With the ink spots at the bottom end, turn over the top 2 cm of the paper and let this end hang over a glass rod. Fasten it with a paper clip.

3. Lodge the glass rod across the top of a 250 cm³ beaker containing water to a depth of 1 cm. The ink spots will be above the level of the water (see Figure 2.13 (b)). It is important that the paper does not sag against the side of the beaker. The purpose of suspending it over the glass rod is to prevent this. Water will travel up the paper to meet the spots.

4. When the water has risen nearly to the paper clip, take the piece of chromatography paper and put it into the oven at 50 °C to dry. Has the pigment separated into different components?

5. Repeat with different coloured felt-tip pens.

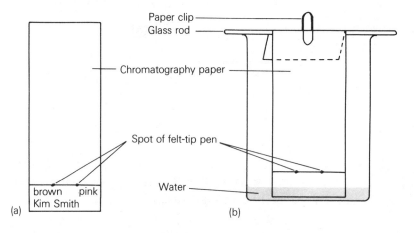

Figure 2.13   Ascending paper chromatography

6. Why is this method described as *ascending paper chromatography*?

Experiment 2.7

To extract the green colouring from green leaves and try to separate it into components

1. Cut up a handful of privet leaves into very small pieces, using scissors. Alternatively, spinach leaves, grass, nettle, or any other green leaves can be used.

2. Put the pieces into a mortar with 3 cm³ methylated spirit and grind with a pestle until you have a really green extract.

3. Put a spot of this extract, using a teat pipette, on to a strip of chromatography paper 10 cm × 2 cm at a distance 2 cm from a short end. When the spot has dried, add another drop in exactly the same place. Apply five drops, one at a time (see Figure 2.14 (a)).

4. Put 5 cm³ methylated spirit into a boiling tube. Hold it in your hand to vaporise the methylated spirit and drive air out of the boiling tube. Stopper the tube with a rubber bung which carries a piece of wire bent into the shape of a hook.

5. Now hook the piece of chromatography paper bearing the green extract. Suspend it inside the boiling tube, with the green spot above the surface of the methylated spirit and the bottom edge of the paper just underneath the surface.

6. Wait until the methylated spirit has risen up the paper to the hook. Take the paper out, and hang it up to dry. Examine your result.

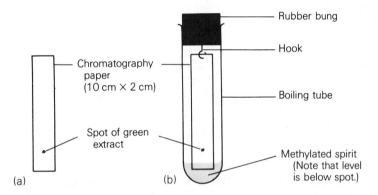

Figure 2.14 Chromatography on extract from green leaves

7. Write a summary of what you have found out. Can you name the green pigment? Explain why water is not a suitable solvent to use.

Experiment 2.8

What interests you?

Would you like to plan an experiment to find out whether you can separate the dyes in Smarties into different colours?

Do you think the colours in rose petals would separate into a number of pigments?

Is the purple colour of blackcurrant juice due to a single substance?

Other coloured substances you could investigate are fabric dyes, food colourings from the kitchen cupboard, or other flower petals and fruit juices. When you have decided which problem you are going to tackle, draw up a plan. Say how you will try to extract the coloured substance and which solvent you will use. Ask your teacher to approve your plan before you start experimenting.

Questions on Chapter 2

1. (a) Describe how you could obtain drinking water from sea water.

   (b) Imagine that you are adrift in a lifeboat with no drinking water left. You do not have laboratory apparatus so you cannot use the method you gave in part (a). You have a primus stove and matches and a kettle and some saucepans. What apparatus can you improvise to supply you with drinking water?

2. (a) What would you do to make the salt in the test tube (see Figure 2.15) dissolve more quickly?

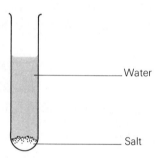

Figure 2.15   Salt and water

   (b) Draw a test tube which contains a mixture of sand and sea water.

   (c) Draw a test tube which contains a mixture of olive oil and vinegar.

   (d) Draw a test tube which contains a mixture of ethanol (alcohol) and water.

3. You are given a white powder and told that it is a mixture of powdered chalk and powdered washing soda. One of the substances is soluble and the other is insoluble in water. Describe what you would do to separate the mixture into a pile of chalk and a pile of washing soda. Say which of the two substances is the soluble one.

4. The Bako Company has patented a red liquid to be used for colouring food. The Kuko Company brings out a new red food dye. The Bako Company suspects that the Kuko dye is the same as their own dye. What can the Bako chemist do to find out whether the Kuko dye is in fact the same as the Bako dye?

5. E_1, E_2 and E_3 are three food colourings. Figure 2.16 shows a chromatogram obtained from spots of E_1, E_2 and E_3. How many different dyes have been used in the manufacture of these three colouring materials?

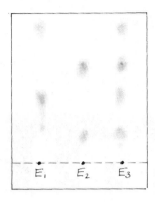

Figure 2.16   A chromatogram

6. Benzoic acid is a white solid. It is soluble in hot water and insoluble in cold water. Describe how you could obtain pure benzoic acid from a mixture of benzoic acid and chalk.

7. ● Water and the solvent 'trichlor' do not mix.

   ● Common salt is soluble in water but insoluble in 'trichlor'.

   ● Wax is insoluble in water but soluble in 'trichlor'.

   ● Sand is insoluble both in water and in 'trichlor'.

   Describe how you could separate a mixture of common salt and wax and sand to obtain a pure sample of each substance.

8. Copy the table, and fill in the spaces.

Summary: Methods of separating mixtures		
Method	*Type of mixture*	*What is the difference between the substances in the mixture?*
Filtration Distillation Fractional distillation Chromatography Separating funnel	Solid and liquid Solute and solvent	

9. Explain the meanings of the following words:
   (a) dissolve   (b) melt   (c) solvent   (d) solute   (e) chromatogram
   (f) filter   (g) filtrate   (h) vaporise   (i) condense   (j) distil

Crossword on Chapter 2

Trace or photocopy this grid (teacher, please see note at the front of the book), and then fill in the answers.

Across

1　An impure form of 8 across (4)
3　Material used in 2 down (5)
7　Method for separating miscible liquids (12)
8　Obtained by evaporating sea-water (4)
11　A change of state will make water into this (5)
12　When 7 across runs _____, we say it is *continuous* (7)
16　_____ long hair in the lab (3)
18　Gas which can easily be liquefied (7)
19　Used with 4 down (6)
20　If you wear a lab coat, your clothes will not get these (6)
22　A pigment obtained from grass by 2 down (11)

Down

1　This forms on iron (4)
2　Method used for separating pigments (14)
4　Used for crushing lumps (6)
5　The purest form of natural water (4)
6　Choose the right one to make a solution (7)
7　Spreads out through a solvent (9)
9　The middle of bath (2)
10　Left behind after 7 across of crude oil (3)
13　Find this greeting in the middle of coil (2)
14　When to give up! (5)
15　Transfer a liquid (4)
17　A help in filtering (6)
21　Obtained by 7 across of crude oil (3)

3 Acids; Bases; Alkalis

3.1 Where You Find Them

Why are British people sometimes called 'limeys'? It is because British sailing ships used to carry lime juice for the crew to drink. The purpose was to keep them free from the painful disease of scurvy. This affects people who cannot get fresh fruit and vegetables. Before the days of refrigeration, ships could not stock fruit and vegetables for a long voyage. Lime juice gave the crew the same protection, and, even on a long voyage, it did not go bad.

The health-promoting substance in fruits is Vitamin C. It is an acid called **ascorbic acid**. It can now be manufactured. Some people believe that large doses of Vitamin C will prevent them from catching a cold.

There is another acid, **citric acid**, present in *citrus* fruits, such as limes, lemons, oranges and grapefruit. It gives these fruits their *acid* taste. Grapes contain **tartaric acid**. Vinegar is a solution of **ethanoic acid**.

Our stomachs contain **hydrochloric acid**. It is needed for the digestion of foods.

Figure 3.1   These contain citric acid and tartaric acid

Acids are not always good for us. Mouth acids cause tooth decay. Where do they come from? A sticky layer called *plaque* forms on teeth. Bacteria in this layer turn food sugars into acids. Acids attack the calcium compounds in teeth. To prevent tooth decay, you have to remove food from the teeth, rub off the layer of plaque, and remove any acid that has been formed. Toothpaste contains powdered chalk and detergent. Brushing removes particles of food, the detergent removes plaque, and chalk removes acids by means of a chemical reaction. Glycerine and water are present in toothpaste to make the mixture into a smooth paste.

All acids contain the element hydrogen. In Chapter 9, you will see how acids can be made to give hydrogen gas in a chemical reaction. The acids you meet in the laboratory are **hydrochloric acid, sulphuric acid** and **nitric acid. Do not taste these acids**. They are stronger than the acids in fruit juices and vinegar.

Bases are the opposite of acids in many chemical reactions. A base will cancel out an acid in a chemical reaction called **neutralisation**. A person who is suffering from acid indigestion may swallow a weak base to neutralise the excess of acid in the stomach.

A soluble base is called an **alkali**. **Sodium hydroxide** is a common alkali. It is often called **caustic soda** (*caustic* means *burning*) because it is a strong alkali. Alkalis have a soapy feel to the skin. The reason for this is that they change the oil in the skin into soap. (You can use an alkali to make soap in Experiment 8.2). Table 3.1 lists some common acids and alkalis.

Figure 3.2   Some common alkalis

Table 3.1   *Some common acids and alkalis*

Name	Where you find it	
Acids		
Citric acid	Fruit juices	
Tartaric acid	Grapes	
Methanoic acid	Ant stings	
Ethanoic acid (Acetic acid)	Vinegar	
Carbonic acid	Fizzy drinks	
Ascorbic acid	Vitamin C	
Sulphuric acid	Car batteries	
Hydrochloric acid	Stomach juice	
Alkalis		
Calcium hydroxide (Its solution is called limewater.)	Used for treating soil which is too acid to be fertile	*continued*

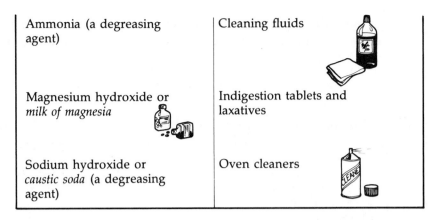

Ammonia (a degreasing agent)	Cleaning fluids
Magnesium hydroxide or *milk of magnesia*	Indigestion tablets and laxatives
Sodium hydroxide or *caustic soda* (a degreasing agent)	Oven cleaners

Only weak acids and alkalis are swallowed or used on the skin. If someone spills a strong acid on their skin, you should not pour a strong alkali on to it. There are two reasons for this. Firstly, heat is given out in neutralisation. Secondly, the strong alkali will burn the skin. The first aid treatment is to wash the skin with plenty of water.

Note the difference between a **concentrated** acid and a **strong** acid. A concentrated solution of acid is one that contains a large amount of acid per litre of water. A dilute solution of an acid contains a small amount of acid per litre of water. A strong acid is one that reacts rapidly with bases and other substances. A weak acid is one that reacts slowly with bases and other substances.

Some bases and alkalis are found in nature. Indians in Peru, South America, suffered from the tropical disease malaria. They found that chewing the bark of the cinchona tree cured their fever. Chemists analysed the bark and found that it contained a base called *quinine*. This is now manufactured by the chemical industry.

It is useful to be able to tell whether a chemical is an acid or an alkali or neither, in which case it is called a **neutral** substance. A substance which will tell you whether a solution is acidic or alkaline or neutral is called an **indicator**. Many indicators can be extracted from plants.

3.2 Extraction of Indicators

Experiment 3.1

To extract an indicator from red cabbage and test it

1. Take two leaves of red cabbage. Tear them into pieces and put them into a 250 cm³ Pyrex beaker.

2. Cover with water. Warm and stir with a glass rod. The water will extract most of the colouring material to become a purple colour, and the leaves will turn almost white.

3. Pour off the extract into a 50 cm³ beaker.

4. Set out three test tubes in a rack. Label them: distilled water, hydrochloric acid and sodium hydroxide solution. Half fill each test tube with the correct solution.

5. Use a teat pipette to add three drops of the coloured extract to each test tube.

6. Make a note of the colour in each test tube. Tabulate your results.

Solution	Distilled water (Neutral)	Hydrochloric acid (Acidic solution)	Sodium hydroxide (Alkaline solution)
Colour of red cabbage indicator			

7. Use the red cabbage indicator to tell you whether solutions of the following substances are acidic, alkaline or neutral: limewater (calcium hydroxide solution), vinegar (ethanoic acid), washing soda (sodium carbonate), 'bicarb of soda' (sodium hydrogencarbonate), common salt (sodium chloride), sulphuric acid and ammonia.

8. See whether you can get indicators from beetroot, blackcurrants, blackberries, elderberries, rose petals, dahlia petals and nasturtium petals. Many blue, red and purple petals and fruits will give results. You can continue this topic at home and test out your indicators on household substances.

3.3 Litmus and Other Indicators

Experiment 3.2

To find the acid, alkaline and neutral colours of indicators

1. Take a rack of clean test tubes. Fill the tubes quarter full with hydrochloric acid, sulphuric acid, distilled water, sodium hydroxide solution, calcium hydroxide solution. Label the test tubes as to their contents.

2. With a teat pipette, add two drops of litmus solution. Record the colour in each solution.

3. Repeat the experiment with the indicators methyl orange and phenolphthalein.

4. Tabulate the colours of the three indicators.

Indicator	Neutral	Acidic	Alkaline
Litmus			
Methyl orange			
Phenolphthalein			

Experiment 3.3

Using universal indicator to find the pH numbers of some solutions

1. Take a rack of clean test tubes. Fill each one half full with one of the solutions listed in the table. Label each test tube.

2. Add three drops of universal indicator to each test tube.

3. Make a copy of the table. Fill in the colours of the indicator.

4. Universal indicator has a colour chart. Each shade on the chart has a number called a pH number. Match the colour of each of your solutions with one of the shades on the universal indicator colour chart. Write down the pH number of each solution.

Chemical	Universal indicator colour	pH number
Sulphuric acid (a strong acid) Ethanoic acid (a weak acid) Sodium hydroxide (a strong alkali) Ammonia solution (a weak alkali) Distilled water (neutral)		

5. Now you can use universal indicator to tell you whether various solutions are strongly acidic or weakly acidic, strongly alkaline or weakly alkaline or neutral. Some suggestions are listed, but there will be others that you want to try. If you are testing a solid, you will have to moisten it first. Some suggestions are:

vinegar	black coffee	limewater
orange juice	health salts	oven cleaner
sugar	milk of magnesia	tartaric acid
baking soda	washing soda	common salt
soap	milk	detergent

6. Make out a table, and classify the substances you have tested.

Acids		Alkalis		Neutral substances
Strong acids	Weak acids	Strong alkalis	Weak alkalis	

7. You may like to display your results on a larger copy of this chart.

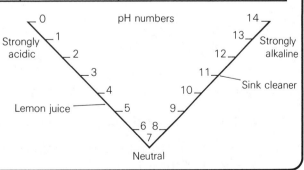

Experiment 3.4

How does universal indicator do it?

How does universal indicator manage to turn different colours in different solutions? Is it a mixture of indicators? You know how to set about separating mixtures: use paper chromatography. Water will not work well for this separation. Some good solvents to use are:

1. 60% butan-1-ol + 20% ethanol + 20% bench ammonia

2. 75% water + 10% ammonium sulphate solution (260 g/l) + 15% ethanol

3. 50% propanone + 50% methanol

Take care; some of these solvents are flammable. Do not use them near a flame. When you have obtained your chromatogram, dry it. How many components have you separated? Test each of the components with (a) hydrochloric acid, (b) sodium hydroxide, (c) ethanoic acid and (d) ammonia.

Write an account of the results of your investigation.

3.4 Neutralisation

Experiment 3.5

To find out what is formed when sodium hydroxide is neutralised by hydrochloric acid

1. With a measuring cylinder, measure 25 cm^3 sodium hydroxide solution into a clean conical flask (see Figure 3.3 (a)).

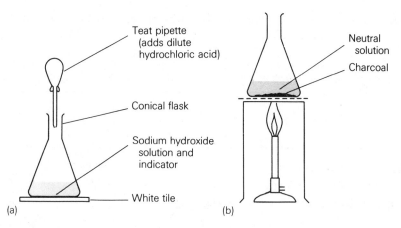

Figure 3.3   Neutralisation of an alkali by an acid

2. Add three drops (*no more*) of screened methyl orange indicator.

3. Take dilute hydrochloric acid from a plastic 50 cm^3 beaker with a teat pipette. Add the acid drop by drop, swirling the conical flask with your

other hand as you do so. Stop when the neutral colour of grey is seen. If the solution goes purple, you have added too much acid and must repeat the experiment.

4. The solution is now neutral, but the indicator must be removed. Add a spatula measure of animal charcoal.

5. Warm and swirl the conical flask for five minutes (see Figure 3.3 (b)).

6. Filter. Evaporate the filtrate to dryness in an evaporating basin.

7. As an exception to our laboratory rules, in this experiment you can taste the product, carefully, on the end of a clean finger.

8. What is the product of the reaction? Copy and complete the word equation

Hydrochloric acid + Sodium hydroxide ⟶ _____ + Water

Experiment 3.6

What happens when you take 'bicarb of soda' for acid indigestion?

'Bicarb of soda' has the proper chemical name of sodium hydrogencarbonate. It is an ingredient of many antacid tablets. The acid which is present in the stomach is hydrochloric acid. It is essential for the digestion of food, but an excess of hydrochloric acid causes pain.

1. Take a test tube and fill it a quarter full with hydrochloric acid.

2. Add 3 drops of universal indicator. Note the colour.

3. Add a spatula measure of sodium hydrogencarbonate. What do you see that tells you a chemical reaction is happening? Add another spatula measure of sodium hydrogencarbonate. Stop adding it when the reaction stops.

4. Note the colour of the universal indicator.

5. Copy this passage, and fill in the blanks.

In hydrochloric acid, the colour of universal indicator is _____. This tells us that the solution is _____. After we added sodium hydrogen-carbonate, the indicator colour changed to _____. This tells us that the solution is _____. There are two reasons why we know that a chemical reaction has happened. One is that when we added sodium hydrogen-carbonate we saw _____ _____ _____. The other is that the indicator colour change tells us that _____ _____ _____. The name of this kind of chemical reaction is _____. Sodium hydrogen-carbonate is known as the indigestion remedy, _____ _____ _____. Acid indigestion is caused by an excess of _____ _____ in the _____. One place where the reaction we have been studying in this experiment happens is _____ _____ _____.

Experiment 3.7

A colourful neutralisation

1. Place 50 cm³ of a saturated solution of boric acid in a 250 cm³ beaker.

2. Add several drops of universal indicator. Stir with a glass rod. Note the colour of the indicator.

3. Take some sodium hydroxide solution which is one quarter of the concentration of bench sodium hydroxide solution. With a teat pipette, add a few drops of sodium hydroxide solution to the beaker. Stir, and note the colour.

4. Continue to add sodium hydroxide solution drop by drop and to stir the solution. Make a note of all the colours you see. Explain why the colour of the indicator is constantly changing.

5. Can you make the seven colours appear again in reverse order?

Experiment 3.8

Some experiments with dyes

Many dyes are obtained from animals and plants. Cochineal and turmeric are used as food colourings, and you are likely to find them in your kitchen. You may like to investigate their behaviour (a) as dyes, (b) as indicators and (c) in chromatography.

1. Make a solution of turmeric by warming it with water or with a mixture of water and ethanol. Filter the solution.

2. Immerse a piece of white cotton cloth for one minute in a hot solution of the dye. Remove the cloth, rinse it, and dry it in warm air.

3. Repeat step 2 with pieces of woollen fabric, nylon fabric and any other materials you want to try.

4. You now have a number of fabrics died with turmeric. Which has taken the dye best? Test the fabric with drops of (a) acid, (b) alkali and (c) household bleach.

5. Extend your work to different dyes and fabrics. Make a table of your results.

Dye: Turmeric				
Fabric	How well does it take the dye?	Action of acid	Action of alkali	Action of bleach
Cotton Nylon				

Questions on Chapter 3

1. In the television programme 'Take Nobody's Word For It' in January 1987, the Prime Minister, Mrs Margaret Thatcher, was interviewed in the kitchen of 10 Downing Street. She was talking about chemistry. The Prime Minister took three glasses containing colourless liquids and poured a red cabbage extract into each one. In the first glass, the cabbage extract was purple. In the second glass, the cabbage extract turned green. In the third glass, the cabbage extract turned red. What ordinary kitchen chemicals do you think the PM had put into the three glasses?

2. Fill the flask! Copy or trace the flask, and fill in the answers to the clues. All the answers read across.

   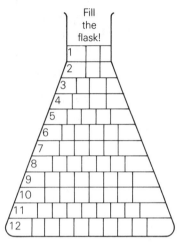

   1,2,3   The colours of litmus in acid and alkali are _____ _____ _____ (3,3,4)

   4   This forms when acid meets alkali (4)

   5   The colour of universal indicator in pure water (5)

   6   The acid which lemons contain (6)

   7   Acid indigestion can be cured by drinking a solution of a weak _____ (6)

   8   Neither acidic nor alkaline (7)

   9   The acid in vinegar (8)

   10   Fizzy drinks contain this acid (8)

   11   Car batteries contain this acid (9)

   12   These tell you whether a solution is acidic or alkaline (10)

3. Someone tells you that the purple dye in elderberries is an indicator. Describe all that you would do to extract the colouring material and test it to find out whether the claim is true.

4. Take home some universal indicator paper. Use it to test foods and cleaning materials and anything else which interests you. Write down a list of acids and alkalis in your home. Classify them into strong acids and weak acids; strong alkalis and weak alkalis.

5. (a) What is the difference between a concentrated solution and a dilute solution?

(b) What is the difference between a weak acid and a strong acid?

(c) How could you tell the difference between a concentrated solution of a weak acid and a dilute solution of a strong acid?

6. (a) Someone spills a concentrated solution of a strong acid on their hand. The first aid treatment is washing with water. Why is the acid not neutralised with a strong alkali?

(b) If a concentrated solution of a strong acid is spilt on the floor, a weak alkali, such as sodium hydrogencarbonate, is used to neutralise it. Why is a strong alkali not used? Why is water not used? How do you know when you have added enough sodium hydrogencarbonate?

(c) A treatment for a bee sting is to dab it with sodium hydrogencarbonate solution. For a wasp sting, vinegar is a remedy. What can you deduce from this about bee stings and wasp stings?

7. Explain the meaning of the following words:

(a) acid (b) base (c) alkali (d) neutralisation (e) indicator

Crossword on Chapter 3

Trace or photocopy this grid (teachers, please see note at the front of the book), and then fill in the answers.

Across

1 A good source of indicator (3,7)
6 Experiments are a good way to _____ about science (5)
7 You find sulphuric acid in this (7)
10 An alkali (6,9)
13 A degreasing agent (7)
15 The sort of solution you get when you add an alkali to an acid (7)
16 Methyl _____ is an indicator (6)
17 This is what the farmer got when he treated his acidic soil in a rainstorm (9)

Down

2 Turn from liquid to vapour (9)
3 An acid in fruits (6)
4 One form of a strong element (4,4)
5, 9down. This tells you whether you have a strong alkali or a weak alkali (9,9)
6 A source of 3 down (5)
8 A capital city in Europe (4)
9 See 5 down
11 Shot with a bow (5)
12 Without carbonic acid, soft drinks are _____ instead of fizzy (4)
14 Average (4)

4 Elements and Compounds

4.1 Metallic and Non-metallic Elements

Metallic

There is gold in the River Madeira, deep in the Amazon jungle in Brazil. Four thousand men are searching for it. Between them, they take £200 million worth of gold from the river bed each year. Rafts anchor on the river. An engine on each raft operates a sort of massive vacuum cleaner. A diver descends 20 metres to the river bed. He carries a 20 centimetre hose attached to the vacuum cleaner. Through it, tonnes of mud containing a few grams of gold dust are sucked up to the raft. Divers spend 6 hours at a stretch under water. Many are boys, 14 or 15 years old. The diver is linked to the raft by an air line as well as the hose. If a diver hits a good vein, other divers muscle in. Down there in the darkness, it is easy for air lines and hoses to become entangled with the wires that anchor the rafts. If his air line is cut, a diver has just enough time to take off his lead-weighted belt and float to the surface for air. If there is a snarl of wires at the bottom, a diver may be trapped. Some bodies emerge downstream; some are never found.

Divers get a 40 per cent share of the gold they find, and there is no shortage of volunteers for the job. The high price of gold is due to its beautiful colour and shine and the fact that it never tarnishes. People have always been prepared to take great risks to find it.

Figure 4.1   Untarnished after 3000 years – the death-mask of Tutankhamen

Gold is an **element**. That is to say that there is no way of splitting it up into simpler substances. **An element is a pure substance which cannot be separated into simpler substances**. Gold is a metal. There are two kinds of elements, metallic and non-metallic. Metallic elements are dense. They can be hammered into different shapes without breaking and pulled out into wire form. Gold is unusual in that it never tarnishes. Most metals are shiny when freshly cut, but react slowly with air to form a dull surface layer. Metals are **sonorous**; that is, they make a ringing noise when struck. They are good conductors of heat and electricity. Other metallic elements are aluminium, calcium, chromium, copper, iron, lead, magnesium, sodium, tin and zinc.

Non-metallic

You are able to swim safely in your neighbourhood pool thanks to chlorine. It is a disinfectant, killing bacteria which would otherwise infect the pool water. Drinking water also contains chlorine, at a much lower concentration. In the days before tap water was chlorinated, people used to catch diseases by drinking water. Chlorine is a non-metallic element. It is a green gas, denser than air, with a choking smell. It is extremely poisonous. In the First World War, chlorine was used as a weapon. In April, 1915, British troops were horrified to see a green cloud rolling towards their trenches. Minutes after it arrived, they were gasping for breath and coughing up blood. Thousands of soldiers were killed and many who survived had permanently damaged lungs. Chlorine is now restricted to killing bacteria.

Other non-metallic elements which are gases are oxygen, nitrogen, neon and hydrogen. Solid non-metallic elements are carbon (charcoal), iodine (used in Experiment 4.2) and sulphur (see Experiment 4.3). Solid non-metallic elements have dull surfaces, apart from diamond and iodine. They are less dense than metals and break when they are hammered. They are poor conductors of heat and electricity.

There are 92 elements found in nature. Most of them fall into the two groups, metallic and non-metallic. Silicon is an exception; it is a **semi-conductor** of electricity, in between metallic elements (conductors) and non-metallic elements (non-conductors). The ability of silicon to conduct electricity can be altered by adding small amounts of chemicals. A silicon chip is a thin slice of a crystal of silicon, 0.5 to 1 cm across. By treating it with chemicals, it is possible to build up a complete electronic circuit, an **integrated circuit**, on a silicon chip. The silicon chip is the device which has made possible the development of the microcomputer industry. Thanks to the silicon chip, micrcomputers are now found in offices, factories, homes, schools, shops and laboratories. Previously, a computer was a massive piece of equipment which occupied several rooms.

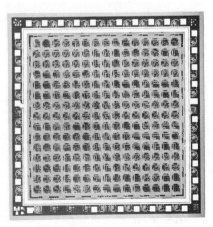

Figure 4.2   A silicon chip (magnified)

Experiment 4.1 will enable you to make observations on a number of elements and try to decide whether they are metallic or non-metallic. The differences are listed in Table 4.1. (You will find a list of elements in Table 6.2.)

Experiment 4.1

To observe the characteristics of a number of elements

1. Make out an enlarged version of the table.

2. Fill in your observations on these elements. Add aluminium, magnesium, lead, sulphur, iodine, copper, iron, calcium and any other elements.

3. Ask your teacher to show you sodium and phosphorus.

4. You will need a circuit to test whether the elements conduct electricity. Connect a battery, a switch and a bulb as shown in Figure 4.3, leaving a gap in the circuit. Attach crocodile clips to the wire on each side of the gap.

Figure 4.3   Circuit for testing conductors

5. Close the gap with, say, a piece of copper, gripping it with the crocodile clips. Depress the switch and notice whether the bulb lights. If it does, a current of electricity must be flowing round the circuit, and the piece of copper closing the gap must be a conductor of electricity.

6. Finally, enter into the last column of the table your opinion as to whether the element is metallic or non-metallic.

Characteristics of some elements

Element	Colour	Surface, shiny or dull	Lump or powder	Can the shape be changed by hammering?	Does it conduct electricity?	Metallic or non-metallic element?
Carbon (charcoal)						
Carbon (graphite)						

Table 4.1   *Characteristics of metallic and non-metallic elements*

Metallic elements	Non-metallic elements
Solids (except mercury)	Some are gases, some are solids; one (bromine) is a liquid
Hard and dense	Most are softer than metals (except diamond) *continued*

Shiny when freshly cut	Dull (except diamond and iodine)
Malleable (can be hammered)	Easily broken when attempts are made to change the shape
Ductile (can be drawn into wire)	
Sonorous (make a pleasing sound when struck)	Not sonorous
Conduct heat	Poor conductors of heat
Conduct electricity	Poor conductors of electricity or insulators (except graphite) or semiconductors (e.g. silicon)

You will have found that there is one exception to the rule that metallic elements conduct electricity and non-metallic elements do not. Carbon is a non-metallic element which exists in different forms. Charcoal is one of its forms, and charcoal is a non-conductor. Graphite is another form of carbon, and yet it conducts electricity. It is shiny in appearance, but it does not have the strength characteristic of metals. As you handle it, flakes rub off on to your hands.

A substance can conduct electricity, and be a metal, without being an element. Brass and steel are metals, but they are not elements. Brass is an alloy, a sort of mixture, of the elements copper and zinc. Stainless steel is an alloy of the elements iron, carbon and chromium. A substance can be a non-conductor of electricity without being an element. The non-conductors, rubber and plastics, for example, are compounds.

Metallic and non-metallic elements are in some senses opposites. The saying that opposites attract one another seems to be true in chemistry. Try these experiments to see what happens when you bring metallic and non-metallic elements into close contact.

4.2 Compounds

Experiment 4.2

To bring together iodine and aluminium

1. Take half a spatula measure of aluminium powder. Put it into a test tube half full of water.

2. Grind some iodine in a mortar with a pestle. Add half a spatula measure to the aluminium. *Do not* grind aluminium and iodine together; this is dangerous.

3. Cork the test tube, and shake it from time to time over the next ten minutes.

4. Can you still see the silvery–grey colour of aluminium? Can you still see the black colour of iodine? Describe what you see. What do you think this substance is?

Experiment 4.3

To heat a mixture of iron and sulphur

1. Take a mixture of iron filings (7 g) and powdered sulphur (4 g). Make sure they are thoroughly mixed. Do two tests on the mixture.

2. Sprinkle some of the mixture on to a piece of paper. Bring a bar magnet underneath the paper and move it towards one end of the paper as shown in Figure 4.4 (a). Repeat.

3. Put a spatula measure of the mixture into a test tube half full of water. With a thumb on top, shake the tube.

4. Make a note of whether (2) and (3) have separated the mixture into iron and sulphur, and, if so, why.

5. Put some of the mixture into a Pyrex test tube until it is one third full. Clamp it at the open end.

6. Heat with a Bunsen burner right at the end of the tube as shown in Figure 4.4 (b). When you notice the mixture glowing red, take away the Bunsen and watch what happens.

7. Allow the tube to cool. Tip out the contents into a mortar. If you cannot obtain the contents any other way, you will have to smash the test tube. You must first wrap it up in a paper towel to prevent bits of broken glass flying about. Grind the contents with a pestle.

8. Repeat steps (2) and (3) to test the contents of the test tube.

9. Record whether (2) and (3) separated iron and sulphur.

10. Observe the appearance of the material you obtained from the test tube. Does it look the same as before heating? Can you see particles of grey iron and yellow sulphur in the mixture?

11. Is observation (9) on the mixture after heating the same as observation (4) on the mixture before heating? If it is different, can you explain why?

12. What did the red glow that spread through the Pyrex tube after you took away the Bunsen burner tell you?

13. What do you think you took out of the test tube?

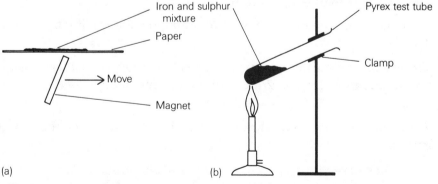

Figure 4.4 (a)   Effect of a magnet on a mixture of iron and sulphur
           (b)   Heating a mixture of iron and sulphur

You have done two experiments on bringing together a metallic element and a non-metallic element. In Experiment 4.2, you saw the silvery-grey colour of aluminium and the black colour of iodine fade and a new yellow solid form. This looked quite different from either element, and was in fact a compound of the two elements called aluminium iodide:

Aluminium + Iodine ⟶ Aluminium iodide

A compound is a pure substance which consists of two or more elements chemically joined together. Aluminium iodide is not a mixture of aluminium and iodine. None of the methods of separation described in Chapter 2 will split it into aluminium and iodine. A more drastic method is needed. If aluminium iodide is melted and an electric current is passed through it, aluminium and iodine are formed.

In Experiment 4.3, you saw that a mixture of iron and sulphur could be separated by two methods: a magnet attracted the iron filings and left the sulphur behind; in water, iron sank and sulphur floated. On heating, a red glow spread through the mixture, suggesting that a chemical reaction was occurring. The product of heating looked different from the mixture. It was a grey solid with no specks of yellow sulphur in it, and it could not be separated as before into iron and sulphur. A chemical reaction had occurred to form a compound called iron sulphide:

Iron + Sulphur ⟶ Iron sulphide

In both these cases, two elements combine in a chemical reaction to form a new substance. The new substance is not a mixture of the two elements. It does not have the appearance or characteristics of the elements; it has a new appearance and new characteristics of its own; it is a compound of the two elements.

The differences between compounds and mixtures are summarised in Table 4.2.

Table 4.2   *The differences between compounds and mixtures*

Mixtures	Compounds
1. A mixture behaves in the same way as the substances present in it behave.	A compound does not behave as its elements do. It has a new set of characteristics.
2. No chemical change takes place when a mixture is made.	A compound is formed from its elements by a chemical reaction. Often, heat is given out or taken in when a compound is formed.
3. A mixture can be prepared by mixing elements in any proportions, for example 1 g of sulphur and 99 g of iron or 99 g of sulphur and 1 g of iron.	A compound always contains its elements in fixed proportions, for example, iron(II) sulphide contains 7 g of iron to 4 g of sulphur.

Questions on Chapter 4

1. Divide the following list into elements and compounds:

gold, gold chloride, aluminium, sodium chloride, copper iodide, sulphur, silicon, silicon oxide

2. (a) Explain the difference between an element and a compound.

(b) Explain the difference between a compound and a mixture.

3. State whether the following are elements or mixtures or compounds:

salt, chlorine, beer, iron filings, steel, lead, silver chloride, charcoal, sea water, brass, copper sulphate, vinegar, rust, sugar, starch

4. What are these metals used for?

(a) silver   (b) lead   (c) aluminium   (d) gold   (e) chlorine
(f) silicon (g) mercury

5. Describe the differences between metallic and non-metallic elements. Name three metallic elements and three non-metallic elements.

Crossword on Chapter 4

Trace or photocopy this grid (teachers, please see note at the front of the book), and then fill in the answers.

Across

1 A small piece of 5 down (4)
3 Simple substances (8)
5 A beautiful metallic element (6)
6 A reactive non-metallic element (6)
8 A sort of cupboard (6)
9 This sergeant major is not royal (1,1)
12 Can be pulled into wire form (7)
13, 11 down   This happens when elements combine (8,8)
14 A cold circle (6)
15 A useful metal (3)
17 A word which describes many metals (5)
18 Metallic sheen can be _____ through corrosion (4)

Down

2 When metals are _____, they look 17 across (8)
3 Can be used to break up compounds (11)
4 Definitely not crazy (4)
5 A semi-conductor (7)
6 Nearly all non-metallic elements are _____ (they don't conduct 3 down) (10)
7 Gold is often found in this form (4)
10 This may be turned into a compound by 13 across, 11 down (7)
11 See 13 across (8)
12 District Attorney (1,1)
16 Put solid waste in here (3)

5 Air

Air is vital to us. The most important thing in the world is getting our next breath. Air keeps us alive by means of a series of complex biochemical reactions. We are going to look at some simpler chemical reactions which take place in air and find out what part air plays in them.

5.1 Heating Elements in Air

Experiment 5.1

To heat metals in air

1. Take a piece of copper in tongs, and hold it in a Bunsen flame (gas half on, air hole half open) for five minutes. Lay down the copper on a heat-proof mat, and note its appearance.

2. Repeat, using magnesium. **Wear safety glasses.** Do not stare at the flame.

3. Put a piece of porcelain on a gauze supported by a tripod. A piece of broken crucible or evaporating basin is suitable. Put a piece of tin on to the porcelain, and heat for five minutes. Observe closely.

4. Repeat, using zinc. Repeat, using lead.

5. Make a table of the metals heated and your observations.

Metal	Appearance	What do you see on heating?	What does the product look like?
Copper Zinc Lead			

6. Have all these metals gained a coating of some substance? When you scrape away the coating, can you see the metal underneath?

7. Does air have to be present before the metals gain a coating? Can you think up a method for finding an answer to this question? Take copper as your example. How can you heat it in the absence of air? There are various possibilities. You could wrap copper up in some other material, some material which will withstand being heated. You could cover it with a layer of some substance to keep out air. Write down a plan for excluding air while you heat a strip of copper. Show the plan to your teacher, and, if he or she approves, carry it out.

After you have done your own experiment, you may like to try Experiment 5.2.

Experiment 5.2

Is air necessary for the formation of a black coating on copper?

1. Take a hard glass test tube. Stopper it, and heat at the closed end steadily at one spot for half a minute. The glass bursts under the pressure of hot gas, and you now have a test tube with a hole in the end to set up as in Figure 5.1. Connect one gas tap to A and connect another to a Bunsen burner at C. Put a piece of copper into the test tube.

2. Turn on the gas at A, in the half-way position. Count five. Standing well back to avoid singeing your hair, light the gas at B. Turn down the gas until the flame is 3 cm high.

3. Light the Bunsen at C and heat for five minutes.

4. Switch off the gas at C. Keep the gas flowing at A until the copper is cool. Then switch off.

5. Note whether a black coating has been formed

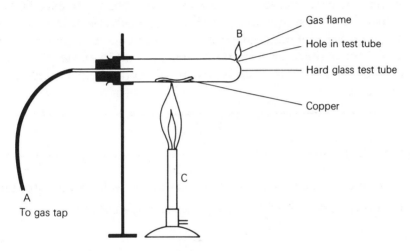

Figure 5.1   Heating copper in the absence of air

6. Now answer the question. Is air necessary for the formation of a black coating on copper?

Experiment 5.3

Is there a change in mass when copper gains a black coating?

1. Fill a crucible one-third full of copper turnings. These must not be greasy as a loss of grease on heating will upset the results.

2. Put the crucible on to a top-loading balance, and read off the mass.

3. Put the crucible into a pipeclay triangle on a tripod (as in Figure 5.2). Heat it for about ten minutes, until the copper turnings look black. Do not use a lid. Cool.

4. Find the mass of the crucible and black copper turnings.

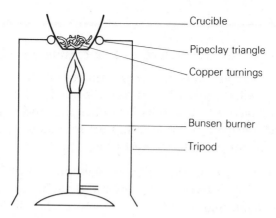

Figure 5.2   Heating copper turnings

5.   Mass of crucible + copper before heating =
                     Mass of crucible + copper after heating =
                                    Difference in mass =

Has there been an increase or a decrease or no change in mass? Compare results with the rest of your class. Have you been able to decide whether copper gives something to or takes something from the air or does neither? Since this is an important turning point in our work on air, you may like to check your result by means of another experiment along the same lines.

Experiment 5.4

Is there a change in mass when magnesium burns to form a white solid?

1. Fill a crucible one third full with magnesium ribbon. Put on the lid.

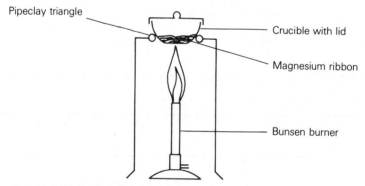

Figure 5.3   Heating magnesium in a crucible

2. With a top-loading balance, find the mass of the crucible and magnesium and lid.

3. Heat the crucible on a pipeclay triangle. From time to time lift the lid with tongs to let air in. The lid prevents particles of white solid from blowing away.

4. When the magnesium no longer glows when you lift the crucible lid, the reaction is over. Switch off the gas, and allow the crucible to cool.

A NEW FIRST CHEMISTRY COURSE

5. Find the mass of the crucible and lid and white solid.

Mass of crucible + lid + magnesium before heating =

Mass of crucible + lid + magnesium after heating =

Difference in mass =

Is there an increase or a decrease or no change in mass?

You will have found that, when air is carefully excluded, copper does not gain a black coat on heating. This opens up three possibilities. When copper gains a black coat, copper

(a) gains something from the air, or

(b) gives something to the air, or

(c) neither gives to nor takes from the air.

A little thought will tell you how to decide between the three possibilities. An experiment will tell you whether there is

(a) an increase in mass, or

(b) a decrease in mass, or

(c) no change in mass when copper gains its black coat.

If you have done Experiments 5.3 and 5.4, you will have found out that there is an increase in mass. This proves that copper takes something from the air to form its black coat, and magnesium takes something from the air when it forms a white solid.

These experiments lead us to think about whether all the air is able to combine with copper and with magnesium or whether only part of the air is used up. Demonstration Experiment 5.5 enables us to find out.

5.2 Experiments to Find Out what Fraction of the Air Will Take Part in Chemical Reactions

Demonstration Experiment 5.5

To find the percentage of air used up when copper gains its black coating

1. Set up the apparatus shown in Figure 5.4. The glass syringes each have a capacity of 100 cm^3. Strings hold the plungers to the barrels in case a plunger should fall out. The syringes are both clamped at the same height and connected by pieces of pressure tubing to a narrow silica tube containing copper turnings. The connections must be airtight. At the beginning of the experiment, have 100 cm^3 of air in one syringe and no air in the other.

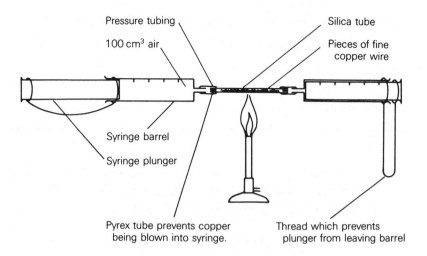

Figure 5.4   Syringe experiment to find out what fraction of the air is reactive

2. Heat the tube containing copper, and drive the air slowly through it from one syringe to the other. Then push the air slowly back. After three minutes, stop heating and allow the apparatus to cool. Measure the volume of air.

3. Repeat the heating, pushing air slowly to and fro from one syringe to the other. Cool again, and measure the volume again.

4. A time will come when the volume of air stops changing. When it has reached a steady value, note the volume of air left in the syringe.

5. Stop heating. When the apparatus is cool, measure the volume of air again.

6. Test the gas in the syringe. Drive it into a test tube by pushing in the plunger. Take a lighted taper, and put it into the test tube. Does the taper continue to burn?

7.   At the beginning of the experiment, volume of air $=$ 100 cm^3

                 At the end of the experiment, volume of air $=$

                      Volume of air used up by copper $=$

           Percentage by volume of air used by copper $=$

8. Is there a change in the appearance of the copper?

    If you weighed the copper before and after heating, what would you notice?

    Why did you have to cool in step 5 before you read the volume of air in the syringe?

Experiment 5.6

Is part of the air used up when a candle burns?

1. Attach a candle to a cork raft by melting the base of the candle. Float it in a trough of water.

2. Light the candle, and put a bell jar over it as shown in Figure 5.5.

3. Stopper the bell jar and wait.

4. The experiment can be done with a gas jar instead of a bell jar.

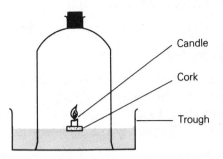

Figure 5.5   Burning a candle in a bell jar

Experiment 5.5 can be repeated with other metals. The result is always the same: 21 cm³ out of every 100 cm³ of air are used by the metal; that is 21 per cent or about one fifth of the air. Four fifths of the air will not react with most metals.

Burning is one of the most important reactions which use air. In Experiment 5.6, you burned a candle. This is not an element but is a mixture of compounds obtained from petroleum oil. When the candle flame went out, you saw water rise up inside the bell jar to take the place of the air which had been used up. The gas left in the bell jar contains gases formed from the burning candle as well as the unreactive part of the air. This is why we cannot use this experiment to tell us exactly what fraction of the air is used up.

Your experiments have shown that air is a mixture of at least two gases. One is a reactive gas which makes up one fifth of the air. The other is an unreactive gas which makes up four fifths of the air. It is named **nitrogen**.

The reactions which we have studied can be explained as combination with oxygen, a process called **oxidation**.

Copper + Oxygen ⟶ Copper oxide

Magnesium + Oxygen ⟶ Magnesium oxide

Combustion is oxidation in which energy is given out. We talk about the combustion of foods in the body; in this type of combustion, there is no flame. In the combustion of paraffin in a stove, there is a flame, and we call this type of combustion **burning. Burning is combustion accompanied by a flame.** For example, magnesium burns in oxygen to form magnesium oxide.

5.3 Preparing Oxygen and Burning Elements in Oxygen

Hydrogen peroxide solution is a colourless liquid which decomposes to give oxygen and water.

$$\text{Hydrogen peroxide} \longrightarrow \text{Oxygen} + \text{Water}$$

The decomposition is slow.

There are substances which **catalyse** the reaction; that is, speed up the reaction. The **catalyst** which you use in Experiment 5.7 is manganese(IV) oxide, a compound of oxygen with the metal manganese. (Chapter 6 will explain what (IV) means.)

Figure 5.6 shows how to test for oxygen. Substances burn much faster in oxygen than in air. A wooden splint which you have lit and then almost blown out bursts into flame in oxygen. **Oxygen relights a glowing splint.**

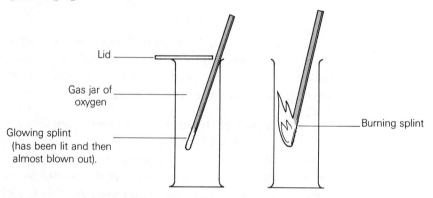

Figure 5.6   Testing for oxygen

Experiment 5.7

To prepare oxygen and to burn some elements in it

1. Assemble the apparatus shown in Figure 5.7. Put a spatula measure of manganese(IV) oxide in the side-arm tube.

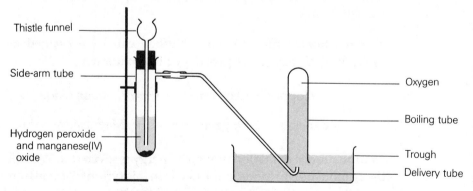

Figure 5.7   Apparatus for preparing oxygen

2. Lay four boiling tubes in the trough so that they fill with water. Hold one over the end of the delivery tube.

3. Add 20 cm^3 of hydrogen peroxide solution to the manganese(IV) oxide.

4. Discard the first tube of gas collected as this is largely air. As soon as the second tube is full, put a cork in the tube under water, and quickly place the open end of the next tube over the delivery tube.

5. Collect and cork four boiling tubes of oxygen and stand them in a rack. Carry out the following tests, with the tubes standing in the rack (**not in your hand**). Do not take out the cork until the burning sample is ready to be put into the tube. After each test, add three drops of litmus solution, cork and shake the tube. **Wear safety glasses**.

6. *Carbon*. Place some powdered carbon in a combustion spoon. Heat to redness, then quickly put the spoon into a boiling tube of oxygen.

7. *Sulphur*. Place some sulphur in a combustion spoon. Heat until the sulphur burns, and quickly transfer to the tube of oxygen.

8. *Iron*. Wrap some iron wool round a combustion spoon. Heat until it is red hot. Put into a tube of oxygen.

9. *Magnesium*. Wrap a short piece of magnesium ribbon round a combustion spoon. Ignite it, and quickly place in a tube of oxygen. Do not stare directly at the flame.

10. Ask your teacher to demonstrate the combustion of sodium, calcium and phosphorus in oxygen.

11. Make a table to show your observations on the colour of the flame, the nature of the product and the action of litmus on a solution of the product.

You will have noticed that oxygen allows the elements to burn more brightly than air does. Oxygen is a better **supporter of combustion** than air is. Your table of results should look like this.

Table 5.1   *Combustion of elements in oxygen*

Element	Metallic or non-metallic	How does it burn?	Appearance of product	Colour of litmus in solution of product	Is the solution acidic or alkaline?	Is the product an acid or a base?
Carbon	Non-metal	Red glow	Colourless gas	Red	Acidic	Acid
Sulphur	Non-metal	Blue flame	Misty gas	Red	Acidic	Acid
Iron	Metal	Yellow sparks	Blue-black solid	Insoluble		
Magnesium	Metal	White light	White solid	Blue	Alkaline	Base
Sodium	Metal	Yellow flame	Yellowish solid	Blue	Alkaline	Base
Calcium	Metal	Red flame	White solid	Blue	Alkaline	Base
Phosphorus	Non-metal	Yellow flame	White solid	Red	Acidic	Acid

When elements burn in oxygen, they combine with oxygen to form **oxides**. A compound of oxygen and one other element is called an **oxide**. The oxides of a number of elements are described in Table 5.1. You can see many examples of the amazing differences between a compound and the elements which combine to make it. The oxides of metals are **bases**. Some bases, for example iron oxide, are insoluble and cannot be tested with litmus. Soluble bases are called **alkalis**. Alkalis are a subset of bases. Bases are compounds which react with acids to give a salt and water only. Most non-metallic elements form acidic oxides. A few non-metallic elements have oxides which are neutral and insoluble. Table 5.2 summarises the information you have obtained.

Table 5.2 *Products of burning elements in oxygen*

Elements

Metallic elements   Non-metallic elements

Burn in oxygen   Burn in oxygen

Oxides of metals,   *Oxides* of non-metallic elements,

Bases   **Acids**

React together
by
Water   Neutralisation

to | form

Some bases   Some bases
do not   dissolve:
dissolve:

Insoluble   Alkalis   Salts   (There are some oxides of non-
bases**   metallic elements which are
                                                   neutral.)

5.4 The Gases Present in Air

There are other gases present in addition to oxygen and nitrogen. The composition of air is shown in Figure 5.8.

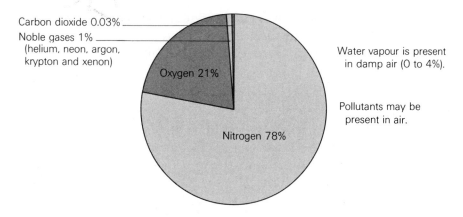

Carbon dioxide 0.03%
Noble gases 1%
(helium, neon, argon,
krypton and xenon)

Water vapour is present
in damp air (0 to 4%).

Pollutants may be
present in air.

Oxygen 21%

Nitrogen 78%

Figure 5.8   The composition of pure, dry air in percentage by volume

Oxygen and Nitrogen From the Air

In Experiment 2.3, we saw how the liquids water and ethanol could be separated by distillation. It is possible, though very difficult, to liquefy air and distil it to give both oxygen and nitrogen. Many gases can be liquefied by cooling and compressing them. This method does not work for air as a very low temperature must be reached before it will liquefy, and it is impossible to find anything cold enough to cool air down so far. The method discovered by Joule and Thompson is used. They found that when a gas is compressed and then suddenly allowed to expand, the expansion results in a cooling. By this method, air can be cooled sufficiently to liquefy it.

Liquid air is a transparent, pale blue liquid. It must be kept below −190 °C in a special type of thermos flask called a *Dewar flask*, which prevents heat from entering. Liquid air is distilled in a carefully insulated fractionating column (see Figure 5.9).

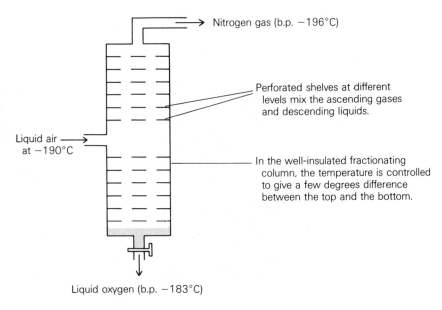

Nitrogen gas (b.p. −196°C)

Perforated shelves at different
levels mix the ascending gases
and descending liquids.

Liquid air →
at −190°C

In the well-insulated fractionating
column, the temperature is controlled
to give a few degrees difference
between the top and the bottom.

Liquid oxygen (b.p. −183°C)

Figure 5.9   Distillation of liquid air

Gradually oxygen accumulates at the bottom as a liquid, and the more easily vaporised nitrogen is taken off from the top as a gas. The boiling points are −196 °C for nitrogen and −183 °C for oxygen.

Oxygen

Pure oxygen is stored under pressure in strong metal cylinders. It has many uses.

Space rockets. In addition to fuel, space rockets also carry their own oxygen for the fuel to burn in, as there is no oxygen in space. The Saturn rockets, which were used to lift American astronauts into orbit for journeys to the moon, carried over 2200 tonnes of liquid oxygen. The first stage, while the jets were roaring, burned 450 tonnes of kerosene in 1800 tonnes of oxygen. Stages two and three were powered by hydrogen burning in oxygen. There was also some oxygen aboard for the astronauts to breathe (see Figure 5.10).

Figure 5.10   Rocket launch

Steel. One tonne of pure oxygen is needed for every 10 tonnes of steel produced from impure iron. The carbon in the impure iron burns away to form carbon dioxide. Modern steel works are usually equipped to make their own oxygen on the site (see Figure 5.11).

Figure 5.11   Steel-making – An oxygen lance injecting the gas into a steel-making furnace

Oxy-acetylene cutting and welding. Acetylene (ethyne) burns more vigorously in pure oxygen than in air, producing a flame with a temperature of about 3000 °C. This temperature will melt iron and steel, which can then be cut or welded (joined together). This is shown in Figure 5.12.

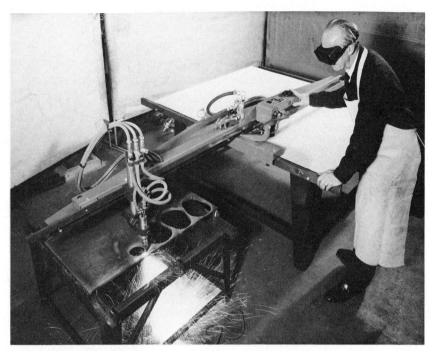

Figure 5.12   An oxy-acetylene flame

Hospitals. In hospitals, oxygen is used to help patients with breathing difficulties, such as pneumonia cases and asthmatics. It is used to revive people who have been dragged out of smoke-filled rooms and people rescued from drowning. It is mixed with anaesthetic gases during surgical operations (see Figure 5.13).

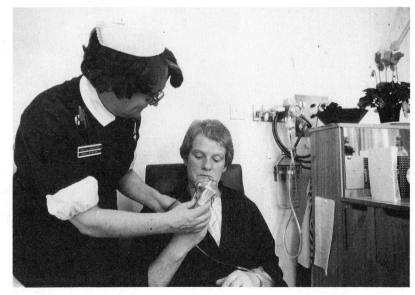

Figure 5.13   Hospitals use oxygen – A patient receiving oxygen therapy

Pollution. Oxygen is used to fight pollution. Figure 5.14 shows the River Protection Service cleaning up the environment by piping oxygen from a tank into polluted river water.

Mountaineers and divers. The air becomes less dense as you climb a mountain. On Mount Everest, which is 8 kilometres high, and on similar peaks, there is not enough air for mountaineers to breathe, and they have to carry cylinders of oxygen on their backs. High-altitude fliers also carry oxygen with them. Underwater explorers may carry a mixture of oxygen and helium in portable cylinders attached to breathing masks, or work inside special capsules, which are supplied with oxygen.

Figure 5.14   Using oxygen to fight pollution

Figure 5.15   Using oxygen in a trout hatchery. Oxygen enters through the pipe on the left of the picture.

Nitrogen

Nitrogen is much less reactive than oxygen. It plays no part in breathing and burning. You would be wrong, however, to suppose that nitrogen is unimportant. Plants need a good supply of nitrogen to keep them alive. Some plants can use the nitrogen in the air. Peas, beans and clover can do this. Inside these plants, nitrogen combines with carbon, hydrogen and oxygen to make proteins. This process is called **fixing** nitrogen. The plants which can **fix** nitrogen have bumps called **nodules** on their roots (see Figure 5.16). The nodules contain **nitrogen-fixing bacteria**. Most plants cannot use nitrogen gas; they rely on obtaining nitrogen compounds from the soil. If the soil is not rich in nitrogen compounds, it yields poor crops. Nitrogen compounds must be added to make the soil more fertile.

Figure 5.16   *Vicia sepium,* a plant with nodules on its roots

Synthetic fertilisers contain nitrates or ammonium salts. These are nitrogen compounds. Nitrates are the salts of nitric acid. They contain nitrogen in a form which all plants can use. Being soluble, nitrates pass easily into the roots of plants. Plants can make proteins from nitrates. Ammonium salts are not used directly by plants. Since they are soluble, ammonium salts penetrate easily into the soil. Soil bacteria turn them into nitrates. These bacteria are called **nitrifying bacteria**.

Some bacteria turn nitrates and ammonium compounds back into nitrogen. These bacteria are called **denitrifying bacteria**. The way in which nitrogen passes from the air into plants and back into the air is shown in Figure 5.17. It is called the **nitrogen cycle**.

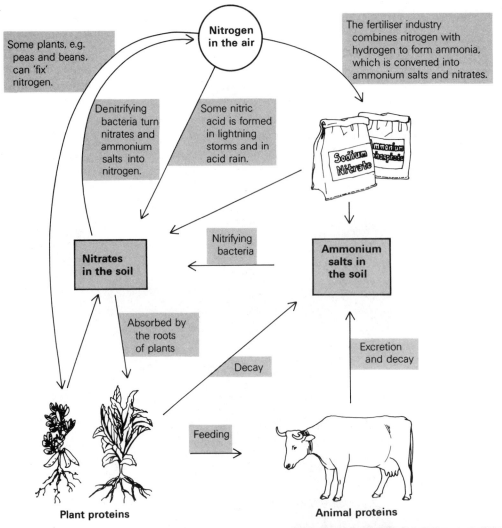

Some plants, e.g. peas and beans, can 'fix' nitrogen.

Nitrogen in the air

The fertiliser industry combines nitrogen with hydrogen to form ammonia, which is converted into ammonium salts and nitrates.

Denitrifying bacteria turn nitrates and ammonium salts into nitrogen.

Some nitric acid is formed in lightning storms and in acid rain.

Sodium Nitrate

Ammonium phosphate

Nitrates in the soil

Nitrifying bacteria

Ammonium salts in the soil

Absorbed by the roots of plants

Decay

Excretion and decay

Feeding

Plant proteins

Animal proteins

Figure 5.17　The nitrogen cycle

The fertiliser industry combines nitrogen with hydrogen to make ammonia. Ammonia is an alkali. It reacts with acids to form ammonium salts. These salts are sold as fertilisers. The fertiliser industry also makes nitrates. Nitric acid can be made from ammonia and then neutralised to give nitrates.

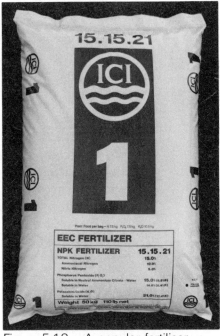

Figure 5.18　A popular fertiliser

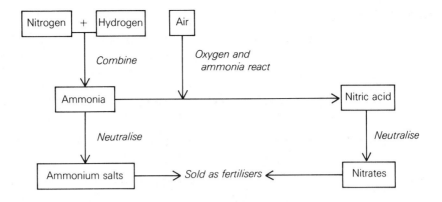

The fertiliser industry sells 7 million tonnes of fertiliser to UK farmers each year. The price is about £120 per tonne. You can see that it is a big business. It enables the British farmer to grow large crops on a small area of land. This is important in a densely populated island. The fertiliser industry also plays a big part in improving very poor agricultural land in Africa and Asia. (You will read more about fertilisers on p. 138.)

Test Yourself on Fertilisers

1. Ammonia is an alkali. What sort of compounds will react with ammonia to form salts? Give the chemical name for this type of reaction. What are the salts of ammonia called?

2. What sort of compounds will react with nitric acid to form salts? Name the type of reaction that takes place. Name the salts of nitric acid.

3. What do plants do with nitrogen? Why do farmers need to put nitrogen compounds on to their fields when there is so much nitrogen in the air?

4. Name a plant which could be sown in poor soil to improve the quality of the soil. Explain how the plant fertilises the land.

5. There are two reasons why nitrates and ammonium salts are used as fertilisers. Explain what these reasons are.

6. Plants use nitrogen from the air. The fertiliser industry takes nitrogen from the air. How is it that the nitrogen content of the air stays at 78 per cent?

Carbon Dioxide

Carbon dioxide is only 0.03 per cent of the air, but it is absolutely essential. Without carbon dioxide, plants could not grow; without plants, animals would starve; and without vegetable crops and meat to eat, we would not survive. You can read about it in Chapter 7.

Water Vapour

The content of water vapour in the air varies, from place to place and from time to time. On an average day in Britain, it might be 2 per cent to 3 per cent.

It is possible to detect the carbon dioxide and water vapour in the air by using these tests:

(1) Carbon dioxide turns a solution called limewater milky (see Chapter 7 p. 95).

(2) Water turns white anhydrous copper sulphate blue (see p. 6).

Noble Gases

The noble gases are a group of gases which are elements and which are chemically unreactive. For a long time, it was believed that they formed no compounds at all, and they were called the *inert gases*. Recently, it was discovered that they can be made to combine with one or two other elements, under very drastic conditions. They are called *helium*, *neon*, *argon*, *krypton* and *xenon*. Helium is very low in density and is used in balloons sent up for meteorological purposes. Neon and argon are used in artificial lights.

Experiment 5.8

To test for carbon dioxide and water vapour in the air

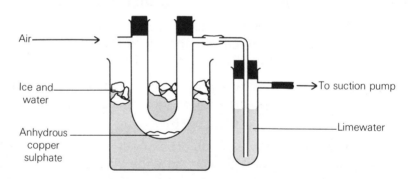

Figure 5.19   Testing for carbon dioxide and water vapour in air

1. Set up the apparatus as shown. The U tube stands in ice-water to condense any water vapour out of the air. It contains anhydrous copper sulphate. *Anhydrous* means *without water*. Anhydrous copper sulphate is a white powder. When it reacts with water, it forms a compound of copper sulphate and water, which is blue. The side-arm tube contains limewater. Limewater is a solution of calcium hydroxide.

2. Using a suction pump, draw a stream of air through the apparatus.

3. Watch to see whether the limewater turns milky.

4. Note whether a liquid condenses inside the U tube, and whether it turns the white anhydrous copper sulphate blue.

5. Copy and complete this sentence:
   Air contains _____, which reacts with limewater to form _____, and also _____, which changes anhydrous copper sulphate from _____ to _____.

5.5 Pollutants

The content of impurities in the air is greater in towns and industrial centres than in the countryside. For a fuller account of pollution, see *Extending Science 1: Air* by E. N. Ramsden and R. E. Lee (ST(P)).

Smoke, Dust and Grit

The air contains millions of tonnes of dust and grit. Some of this solid matter enters the air from natural sources, such as volcanic eruptions and dust storms. Most of it comes from factory chimneys, coal-burning electricity power stations, mines and the exhausts of motor vehicles. The solid particles fall to earth and settle on buildings, pavements and plants. We breathe in these tiny particles. Some of them are bad for our health. The amount of solid matter in the atmosphere is increasing.

Carbon Monoxide

Carbon monoxide is a poisonous gas. Every year 350 million tonnes are sent into the air. The chief culprit is the petrol engine. The exhaust gases from motor vehicles contain some carbon monoxide. (You will read more about this in Chapter 7.)

Carbon monoxide combines with the red pigment in the blood, haemoglobin. It prevents haemoglobin from combining with oxygen, and carrying oxygen around the body. If the level of carbon monoxide reaches one per cent of air, it will kill quickly. At lower levels, it causes dizziness and headaches. At quite low levels, it slows down a person's reactions. Carbon monoxide is most likely to build up in crowded streets at rush hours. It can affect drivers' reaction times when they most need quick reflexes.

Oxides of Nitrogen

Fuels are burned in furnaces and motor vehicle engines. At the high temperatures inside furnaces and engines, some of the nitrogen in the air combines with oxygen. Nitrogen dioxide is formed. This gas reacts with oxygen and water to form nitric acid. You can read about the consequences under *Acid rain*.

Sulphur Dioxide

Sulphur dioxide is a poisonous gas with a penetrating smell. Almost all the sulphur dioxide in the air comes from industrial sources. The chief ones are:

- Many metals are obtained from sulphides (compounds of metals with sulphur). Obtaining these metals from sulphides puts tonnes of sulphur dioxide into the air every day.
- Coal contains sulphur. When coal is burnt, the sulphur in it burns to form sulphur dioxide. This happens in domestic fires, factories and coal-burning electricity power stations.
- Oil contains sulphur. Factories and power stations and homes which burn oil send sulphur dioxide into the air.

Sulphur dioxide is a very irritating gas. It causes coughing and chest pains. Anyone who is already suffering from bronchitis or a lung disease can become very ill by breathing sulphur dioxide. Factories and power stations use tall chimneys to take sulphur dioxide up into the atmosphere. But that is not the end of it, as you will see.

Acid Rain

You have read that metal works, power stations and factories send sulphur dioxide into the air. Motor vehicles send nitrogen dioxide into the air. These gases are both acidic. They combine with water vapour in the air and oxygen to form sulphuric acid and nitric acid. These come down to earth as **acid rain**; see Figure 5.20. The acid rain may fall hundreds of miles from the power station which caused it.

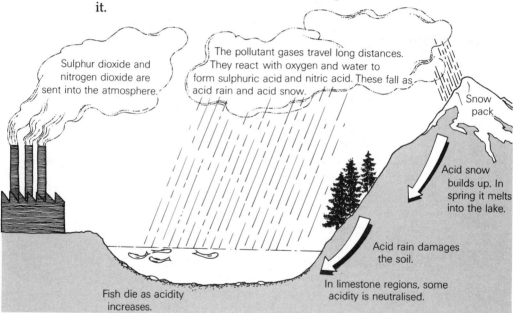

Figure 5.20   Acid rain

Unpolluted rain water is slightly acidic because carbon dioxide from the air dissolves in it to form the weak acid, carbonic acid. Rainwater with a pH less than 5.6 (for pH see p. 35) is described as *acid rain*. It does a great deal of damage.

Damage to Buildings

Acid rain and the acidic gases in the air react with metals, limestone, concrete and cement. All these are building materials. The damage done by acid rain to buildings is widespread:

- The Taj Mahal in India was finished in 1640. For centuries the marble remained as new. Now it is being attacked by acid rain.

- The Acropolis in Greece has stood for 2000 years. Now the polluted air of Athens is corroding away the marble.

- Cologne Cathedral in Germany is decorated with intricate stone statues. Now acid rain and polluted air are eating into the stonework.

- In Holland, the Royal Palace is becoming corroded. Statues on the outside of 450-year-old St John's Cathedral are 'melting away like lollipops' according to one architect.

- In London, every major historic building is affected. At Westminster Abbey and the Tower of London, stonework is being corroded. At St Paul's Cathedral, up to 3 cm of limestone has been eaten away in places.

Damage to Trees

There are dead forests in Czechoslovakia, Poland and Germany. In the famous Black Forest of Germany, 50 per cent of the trees are damaged. Germany has reckoned a loss of £500 million a year in its income from forestry. Acid rain is believed to be the cause of the damage to trees. Here again, pollution costs money.

As well as forests, farmland is affected. Acid rain washes nutrients out of the soil, and crops are robbed of nourishment. West Germany has reckoned that the reduced crop yields cost the country £400 million a year.

How does the damage arise? At first, the nitrates contained in acid rain fertilise the soil and stimulate the growth of plants. Most soils can neutralise a small amount of acid, but they cannot deal with large amounts. After a while, the ability of the soil to neutralise acid is exhausted. Acid rain then attacks the metal salts on which plants depend for nourishment. Acid rain converts calcium, magnesium and aluminium compounds in the soil into soluble nitrates and sulphates. These soluble salts wash out of the soil into the subsoil, where the roots cannot reach them. When this happens, plants begin to suffer from a lack of nutrients. One of the soluble salts formed by acid rain is aluminium sulphate. This salt damages the roots of trees. Damaged roots can be invaded by harmful viruses and bacteria. The trees die slowly as a result of starvation (through lack of nutrients) and disease (through damaged roots).

Damage to Lakes

In Sweden, 4000 lakes are dead, and in 9000 there are few fish left. In Norway, 2000 lakes are dead. In the province of Ontario in Canada, fish have gone from 3000 lakes. The reason why the fish die is that aluminium sulphate is washed into the lake water. Aluminium sulphate comes from the action of acid rain on aluminium compounds in the soil.

Fish pass water over their gills. Aluminium hydroxide is deposited on the gills. As the gills become unable to work, the fish struggle for breath and eventually die. It is not only the fish that die. An acid lake is crystal clear. Plankton, insects, snails, mussels and other forms of life have also perished.

In 1982, Britain began adding lime (calcium hydroxide) to rivers and lakes in south-west Scotland. The aim is to neutralise acidic water and revive stocks of fish. In Loch Fashally, the average pH of rain fell from 5.2 in 1962 to 4.2 in 1975. A fall of one pH unit means a ten-fold increase in acidity. At Pitlochry, a Ministry of Agriculture station measured the pH of rain from one thunderstorm as 2.4, which is more acidic than vinegar, and is the most acidic rain ever recorded. The Lake District and parts of Scotland have rain as acidic as anywhere in the world.

The Welsh Water Authority believes that acid rain has wiped out life in nine lakes and threatens the fishing industry. The authority is pouring tonnes of powdered limestone into lakes which are threatened.

Some lakes are affected more than others:

- One factor is the type of land surrounding the lake. In limestone regions, limestone will neutralise some of the acid rain and keep damage to a minimum.

- Another factor is the rate at which acid is released. In Norway and Sweden, the cold winters mean that they receive acid snow. In the spring thaw, several months' accumulation of acid snow melts and washes quickly into the streams and lakes. There is little time for any neutralisation to take place.

What Can Be Done about Acid Rain?

In 1979, 33 countries signed the Geneva Convention on Long Range Transboundary Atmospheric Pollution. They promised to reduce the amount of pollution they export to other countries. So far, little progress has been made. It costs money to clean up, and experts are arguing about the cost.

Coal-burning and oil-burning power stations give rise to 60 per cent of the sulphur dioxide in the atmosphere. There are ways in which this pollution can be reduced.

- Low-sulphur fuels can be used. Modern oil refineries can produce more low-sulphur oil. Many types of coal can be crushed and cleaned to remove some of the sulphur content.

- Sulphur can be removed while the fuel burns. A new type of burner has been designed. Powdered limestone is injected into the burner while the fuel burns. This removes 80 per cent of both sulphur dioxide and nitrogen dioxide. This type of burner can be fitted when new power stations are built. It cannot be fitted on to old power stations. This method is called *pulverised fluidised bed combustion, PFBC. (Pulverised* means the coal is ground; *fluidised bed* means that the gases which pass through the bed of coal and limestone keep it in motion.)

- Sulphur dioxide can be removed after the fuel has been burnt but before the gases leave the stack. The gases are washed with an alkaline solution. This removes up to 95 per cent of the sulphur dioxide from the flue gases. Equipment for treating flue gases can be fitted to existing furnaces. This method is called *flue gas desulphurisation, FGD*.

In 1986, the UK joined the '30 per cent club'. This is a group of nations who have agreed to cut their emission of sulphur dioxide by 30 per cent. The Central Electricity Generating Board has decided to modify its power stations at Drax in Yorkshire and Fiddler's Ferry in Cheshire. The CEGB is building FGD plants on to the power stations. These will remove 90 per cent of the sulphur dioxide formed.

The Problem of Acid Rain

1. Every year since 1947, Oslo, the capital city of Norway, has sent a Christmas tree to London. This gift is made in memory of Britain's help to Norway during the Second World War. In 1985, the Nature and Youth Society of Norway asked the Oslo city council not to send a tree. The society threatened to 'kidnap' the tree if one was sent. The reason why the group made their protest was that they feel there may not be any Christmas trees in Norway in 40 or 50 years' time. They blame Britain for the damage to Norwegian trees by acid rain. In 1985, Britain had not yet joined the so-called '30 per cent club' of nations.

   (a) Where does the Norwegian tree stand every year?

   (b) Why do the Norwegian Nature and Youth Society think that British pollution can cause damage in Norway?

   (c) The Oslo city council sent the tree anyway. Why do you think they did this?

   (d) What do you think the British government should do to find out where the pollution from British factories and power stations is going?

   (e) In which other countries are trees damaged by pollution?

   (f) What is the name given to this form of pollution?

   (g) Explain how it affects the health of trees.

2. Name

   (a) three pollutants which come from power stations

   (b) three pollutants which come from vehicle exhausts.

3. How does oxygen travel from your lungs to your body tissues? Which pollutant can upset this process?

4. (a) List three ways in which sulphur dioxide gets into the air.

   (b) Describe three kinds of damage which result from the presence of sulphur dioxide in air.

5. Describe three ways in which acid rain costs money.

6. Explain how the sulphuric acid present in acid rain
   (a) damages trees and
   (b) kills fish.

7. Explain why the problem of acid rain is worse in Norway and Sweden than it is in Britain.

8. Removing the causes of acid rain will cost money. How will this money need to be spent?

5.6 The Rusting of Iron

We began our study of air by looking at metals reacting with air. A very important reaction of a metal in air is the rusting of iron. Experiments 5.9–5.12 are designed to allow you to find out several things about iron rusting.

Does iron increase in mass on rusting?

Does iron use up part of the air when it rusts?

Does iron use up the same fraction of the air as the metals which burn in air?

Does iron need both water and air to rust?

Do any other substances need to be present for iron to rust?

What chemical reaction takes place when iron rusts?

What is the chemical name for rust?

Experiment 5.9

Is there a change in mass when iron rusts?

1. Put some dry iron filings into a petri dish. Put the dish on to a top-loading balance, and read off its mass.

2. Add water carefully with a teat pipette so that the filings are wet but none are washed away.

3. Put the petri dish on a window sill and leave it for a week. The water should evaporate. If necessary, put the petri dish and iron filings into the oven to finish drying.

4. Since the water you added has now evaporated, the petri dish and iron filings should have the same mass as before. Reweigh to find out whether this is so.

5. You now know:

   Mass of petri dish + iron filings in (1) = m_1 grams

   Mass of petri dish + iron filings in (4) = m_2 grams

   Are the figures m_1 and m_2 the same? If not, has the mass increased or decreased? What change do you notice in the appearance of the iron filings?

6. Suggest an explanation of your results.

Experiment 5.10

Does iron combine with air on rusting?

1. Take a 100 cm³ graduated tube. To make the volume of air inside the tube 100 cm³, add some water and then turn the tube open side down with your thumb over the end. Carefully let water trickle out until you have 100 cm³ of air. Invert the tube in a deep vessel, and read the volume of air with the water levels inside and outside the tube equal, as shown in Figure 5.21 (a).

2. With your thumb over the open end, turn the graduated tube open end upwards, and push a wad of iron wool to the end of the tube. Carefully invert in a beaker of water.

3. Leave it to stand for a week (see Figure 5.21 (b)).

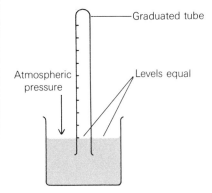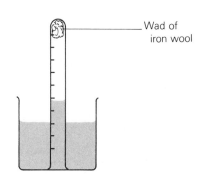

(a) Ensuring the tube contains 100 cm³ of air   (b) Iron rusting in graduated tube

Figure 5.21   Experiment to find what fraction of air combines with iron

4. Put your thumb over the open end, and then open the graduated tube in a deep vessel, so that you can read the volume of air at atmospheric pressure, with the water levels inside and outside the tube equal. Note the volume.

5. Test the gas in the graduated tube with a lighted splint.

6. What change do you see in the iron?
    What does step 5 tell you?
    Can you write a word equation for rusting?

7. Record your results

$$\text{Initial volume of air} = 100 \text{ cm}^3$$
$$\text{Final volume of air} = v \text{ cm}^3$$
$$\text{Volume of air used up} = (100 - v) \text{ cm}^3$$
$$\text{Percentage of air used up} = (100 - v)\%$$

8. What does rusting have in common with breathing and burning?

Experiment 5.11

Can you speed up rusting?

The rusting of steel in Experiment 5.10 is slow. Would you like to see whether you can speed it up? Here are some suggestions to try.

1. Soak the iron wool in bench ethanoic acid for one minute before you start. Then repeat Experiment 5.10. You can use a graduated tube or a measuring cylinder.

2. Other solutions to try are (a) sodium hydroxide solution, (b) phosphoric acid solution and (c) bench hydrochloric acid.

3. Repeat one of the tests in which steel wool rusts quickly in a more precise way. Measure the volume of air in the graduated tube every five minutes. On a piece of graph paper, plot the volume of air against the time.

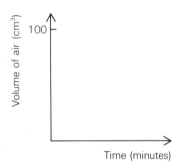

4. What does the shape of the graph tell you?

    You do not really *want* to make iron and steel rust quickly. What use can you make of your results?

    Can you think up an experiment to compare the rates at which different brands of steel wool rust?

Experiment 5.12

To investigate the conditions necessary for iron to rust

1. Set up conical flasks (or specimen tubes or test tubes) containing iron nails under different conditions, as shown in Figure 5.22.

2. To prepare air-free water for experiment (4), put distilled water into the flask, and boil gently for ten minutes. Add some nails, and boil again for five minutes. The nails must remain under the surface. Cool the flask quickly under the cold tap, and pour in a layer of liquid paraffin 1 cm deep. Close the flask with a rubber bung. Boiling drives out air, and the layer of oil keeps out air.

3. Allow the flasks to stand. You will be able to see some differences after two weeks, but if you can leave them for six months the results will be more interesting.

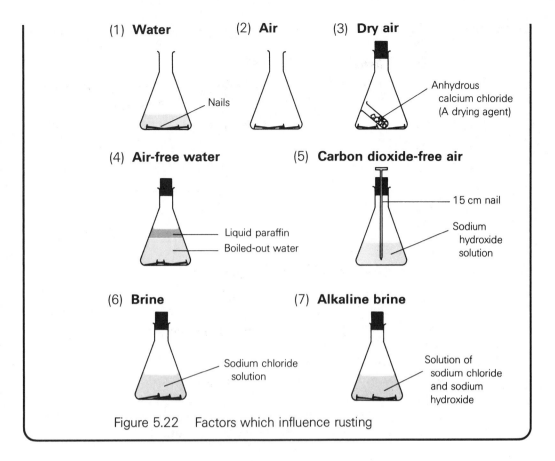

(1) **Water**   (2) **Air**   (3) **Dry air**

Nails

Anhydrous calcium chloride (A drying agent)

(4) **Air-free water**   (5) **Carbon dioxide-free air**

Liquid paraffin
Boiled-out water

15 cm nail

Sodium hydroxide solution

(6) **Brine**   (7) **Alkaline brine**

Sodium chloride solution

Solution of sodium chloride and sodium hydroxide

Figure 5.22   Factors which influence rusting

What Have You Found Out About Rusting?

After the experiments you have done, you should be able to answer these questions about rusting.

1. When iron rusts, does it increase or decrease or stay the same in mass? (See Experiment 5.9.)

2. Does air take part in rusting? (See Experiment 5.10.)

3. What percentage of the air takes part in rusting? (See Experiment 5.10.)

4. Will the air that remains allow a splint to burn?

5. Which part of the air has been used up in rusting?

6. What is rust? Is it a compound of iron and another element? What chemical name would you give to rust?

   Complete the word equation:

   Iron + _____ ⟶ _____

7. Explain what rusting has in common with breathing and burning.

8. As you know from Experiment 5.12, nails rust quickly in water. What happens to nails in (a) dry air, (b) air-free water, (c) alkaline solution and (d) brine?

9. Why is it especially important to find out whether iron rusts in brine (sea water)?

Experiment 5.13

What can be done to prevent iron from rusting?

1. See whether you can think of some methods of protecting iron from rusting. How can you keep air and water away from iron?

2. Decide how you can test the treated iron nails to see how long they last without rusting.

3. Check your plan with your teacher. If he or she approves, start work!

Methods of Preventing Iron from Rusting

Since so many of the tools and machines we use are made of iron or steel, which is an alloy of iron with other metals, it is important to think of ways of preventing iron from rusting. There are various methods.

Paint. Painting is the method used for large iron structures such as bridges and ships. The paint must be kept in good condition. If it chips and exposes the iron beneath, rusting will begin and will spread out underneath the remaining paint.

Oil or grease. A thin film of oil or grease will keep out air and water. This is suitable for moving parts of machinery, for example a bicycle chain.

Metal film. A thin film of another metal can protect iron from attack. Galvanised iron is iron which has been dipped into molten zinc and then lifted out. The layer of zinc protects the iron even if it becomes chipped. Galvanised iron is used for roofing and for dustbins. Food cans are made from iron coated with tin. If the tin coating becomes chipped, the iron rusts. Chromium plating is a process which is both protective and decorative and it is done by an electrical method. Bicycle handlebars are treated in this way. Silver plating is a very expensive way of protecting iron and steel. It is done by electro-plating and is used for decorative objects like vases and candlesticks, and for cutlery.

Stainless steel. Steels are alloys of iron containing some carbon with another metal. The choice of the other metal depends on what kind of steel is required. Some metals make a steel hard-wearing; some metals give a steel which can be sharpened to a knife edge. Stainless steel, which does not rust, is an alloy of iron, carbon and chromium. Air reacts with it to form a thin layer of chromium oxide, which protects the steel from any further attacks by air or water. Stainless steel is more expensive than iron and is only used when rusting would be disastrous. Cutlery is made of stainless steel.

Figure 5.23   This car body has been rust-proofed

Figure 5.24   Galvanised iron girders have been used in the construction of this oil rig

Questions on Chapter 5

1. Explain the reasons for the following statements:

   (a) If air is bubbled through limewater for several minutes, the limewater turns milky.

   (b) When copper is heated in air, there is an increase in mass.

   (c) If iron nails are covered in a sealed tube with tap water they rust, but if the water is first boiled they do not rust.

   (d) A cold dish held in the yellow Bunsen flame soon becomes coated with a black surface.

2. Describe what you see happen when:

   (a) A strip of magnesium ribbon is set alight and then lowered into a jar of oxygen.

   (b) A combustion spoon is filled with sulphur, which is set alight and lowered into a jar of oxygen.

3. Name the three most abundant gases in the atmosphere and two gases which pollute the air.

4. Describe an experiment for finding the percentage by volume of oxygen in the air. What volume of nitrogen can you obtain from $250\,cm^3$ of air?

5.

Sulphur	Carbon	Oxygen
Sodium	Magnesium	Chlorine
Hydrogen	Nitrogen	Calcium

   From this list of elements, choose answers to the following questions:

   (a) Which two elements combine to form an edible substance?

   (b) Which elements are metals?

   (c) Which elements are non-metals?

   (d) Which two elements burn in air to produce oxides which dissolve in water to form acidic solutions?

   (e) Which element has to be kept out of contact with air?

   (f) Which two elements form an explosive mixture?

6. You are asked to find out how well paint and grease and a layer of zinc protect iron from rusting. You are given iron nails, galvanised iron nails, Vaseline, metal paint, water and salt.

   Say what experiments you would do. Use diagrams to illustrate your account.

7. You are given three closed gas jars. Describe three tests you would do to find out whether the gas which they contain is oxygen.

8. When you look at your bicycle, you notice that different parts have received different treatments to prevent rusting. List these different treatments, and explain why different parts of the bicycle need to be treated in different ways.

9. Explain the meanings of these words.

(a) breathing   (b) respiration   (c) combustion   (d) burning
(e) oxide   (f) the nitrogen cycle   (g) pollution   (h) acid rain
(i) rusting

Crossword on Chapter 5

Trace or photocopy this grid (teachers, please see note at the front of the book), and then fill in the answers.

Across

1 Burning this puts pollutants into the air (4)
2 They use oxygen in these (9)
5 A metal which is used in flares (9)
10 This pollutant causes acid rain (7,7)
11 These are polluted by acid rain (5)
13 This will rust slowly (4)
14 See 7 down
15 This low-density gas is used in balloons (6)
16 Particles of dust are this size (4)

Down

1 Burning without a flame (10)
3 Not false (4)
4 A cure for pollution is _____ 10 across before it leaves the factory chimney (8)
6 Four fifths of the air (8)
7, 14 across. This alloy will not rust (9,5)
8 They are damaged by acid rain (5)
9 Coat with zinc (9)
12 The air at the seaside is this (5)

6 Symbols, Formulas and Equations

6.1 Dalton's Atomic Theory

You read about the Atomic Theory in Chapter 1. In this chapter, we shall take the idea of atoms further. Dalton's Atomic Theory can be stated in three parts.

- Matter is composed of enormous numbers of minute particles called atoms. Dalton said that atoms cannot be created or destroyed or split. Since Dalton's day, scientists have succeeded in splitting the atom. There is an enormous release of energy when atoms are split; this *atomic energy* is used to generate electricity in power stations.

- Each element is made up of one kind of atom. They differ from the atoms of all other elements.

- When elements combine to form compounds, the numbers of atoms of the different elements which combine are in a simple ratio to one another. For example, one atom of oxygen combines with two atoms of hydrogen to form a molecule of water. All the molecules of a compound are alike.

It may be useful to list the meanings of some of these terms:

Element. An element is a substance which cannot be broken down into simpler substances.

Compound. A compound is a pure substance which contains two or more elements chemically combined.

Atom. An atom is the smallest particle of an element which can take part in a chemical reaction.

Many elements exist as more complex particles, consisting of a number of atoms. These particles are called molecules. Figure 6.1 shows models of molecules of helium (1 atom), oxygen (2 atoms), phosphorus (4 atoms) and sulphur (8 atoms). Compounds also consist of molecules. Figure 6.2 shows models of molecules of hydrogen chloride, water and methane (which you know as North Sea gas). We can therefore define a molecule in this way:

Molecule. A molecule is the smallest particle of an element or a compound which can exist independently.

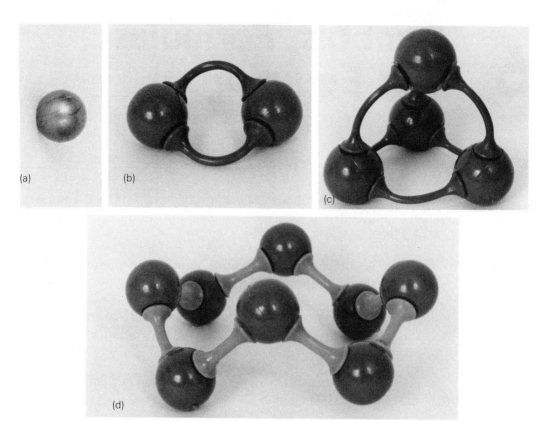

Figure 6.1 Models of molecules of (a) helium, (b) oxygen, (c) phosphorus and (d) sulphur

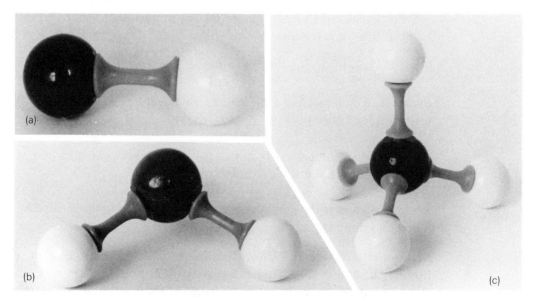

Figure 6.2 Models of molecules of (a) hydrogen chloride, (b) water and (c) methane

6.2 Symbols and Formulas

Chemists use the initial letter of an element to stand for an atom of the element, e.g. C stands for one atom of carbon. Sometimes it is necessary to use two letters: Cl stands for an atom of chlorine, Co for an atom of cobalt, Cr for an atom of chromium. We call these letters symbols. **The symbol of an element is one or two letters which stand for one atom of the element.** Sometimes, the symbols are taken from the Latin names: Pb is the symbol for lead (plumbum), and Cu is the symbol for copper (cuprum). A list of symbols is given in Table 6.1.

A symbol stands for one atom of an element. To represent two atoms, you write a large figure 2 in front of the symbol: 2Cu means two atoms of copper. When a symbol is followed by a small number, e.g. O_2, it shows the number of atoms of the element present in one molecule. H_2 stands for a molecule of hydrogen, consisting of two atoms. S_8 stands for a molecule of sulphur. NH_3 stands for a molecule of ammonia, one nitrogen atom and three hydrogen atoms. It is the **formula** of ammonia.

Sometimes, a group of atoms will go through a series of chemical reactions intact. The hydroxide group, OH, occurs in many compounds. The sulphate group occurs in sulphuric acid and in all sulphates. It has the formula SO_4. Compounds and groups have formulas. A formula consists of symbols and numbers. The symbols show which elements are present in the compound or group. The numbers show the ratio of atoms of each element present. The formula CuO tells you that in copper oxide there is one oxygen atom for every copper atom. The formula H_2SO_4 tells us that in sulphuric acid there are two hydrogen atoms and four oxygen atoms for every sulphur atom. The formula of calcium hydroxide is $Ca(OH)_2$. The two after the bracket multiplies all the atoms in the bracket: there are two atoms of hydrogen, two atoms of oxygen and one atom of calcium. We write the formula of lead nitrate $Pb(NO_3)_2$ instead of PbN_2O_6 because it shows that the compound is a nitrate and will behave similarly to other nitrates, all of which possess the group NO_3.

Table 6.1 shows the formulas of some groups, and Table 6.2 shows the formulas of some compounds.

Table 6.1 *Valencies of elements and groups*

The symbols and valencies of some elements			
	Element	Symbol	Valency
(a) *Metals*	Sodium	Na	1
	Potassium	K	1
	Silver	Ag	1
	Gold	Au	1 and 3
	Barium	Ba	2
	Calcium	Ca	2
	Zinc	Zn	2
	Copper	Cu	2
	Iron	Fe	2 and 3
	Lead	Pb	2
	Magnesium	Mg	2
	Aluminium	Al	3

continued

	Element	Symbol	Valency
(b) *Non-metallic* *elements*	Hydrogen	H	1
	Chlorine	Cl	1
	Bromine	Br	1
	Iodine	I	1
	Oxygen	O	2
	Sulphur	S	2, 4 and 6
	Nitrogen	N	3 and 5
	Phosphorus	P	3 and 5
	Carbon	C	4

The formulas and valencies of some common groups

	Group	Formula	Valency
(a) *'Metallic' group*	Ammonium	NH_4	1
(b) *Non-metallic groups*	Hydroxide	OH	1
	Nitrate	NO_3	1
	Sulphate	SO_4	2
	Carbonate	CO_3	2
	Hydrogencarbonate	HCO_3	1

Table 6.2 *The formulas of some compounds*

H_2O	Water	CO	Carbon monoxide
NaOH	Sodium hydroxide (caustic soda)	SO_2	Sulphur dioxide
		H_2S	Hydrogen sulphide
$Ca(OH)_2$	Calcium hydroxide (solution is limewater)	NH_3	Ammonia
		NH_4Cl	Ammonium chloride
		NH_4NO_3	Ammonium nitrate
HCl	Hydrogen chloride (solution is hydrochloric acid)	$(NH_4)_2SO_4$	Ammonium sulphate
		$CaCO_3$	Calcium carbonate
		$MgCO_3$	Magnesium carbonate
HNO_3	Nitric acid	$Ca(HCO_3)_2$	Calcium hydrogencarbonate
H_2SO_4	Sulphuric acid		
NaCl	Sodium chloride	$Mg(HCO_3)_2$	Magnesium hydrogencarbonate
$NaNO_3$	Sodium nitrate		
Na_2SO_4	Sodium sulphate	Na_2CO_3	Sodium carbonate
$CuCl_2$	Copper chloride	$NaHCO_3$	Sodium hydrogen-carbonate (baking soda)
$Cu(NO_3)_2$	Copper nitrate		
$CuSO_4$	Copper sulphate		
CuO	Copper oxide	CaO	Calcium oxide (quicklime)
$Cu(OH)_2$	Copper hydroxide		
$AlCl_3$	Aluminium chloride	$ZnCl_2$	Zinc chloride
$Al(OH)_3$	Aluminium hydroxide	$Zn(OH)_2$	Zinc hydroxide
Al_2O_3	Aluminium oxide	ZnO	Zinc oxide
$Al_2(SO_4)_3$	Aluminium sulphate	$ZnSO_4$	Zinc sulphate
$CuSO_4.$ $5H_2O$	Copper sulphate crystals	$Na_2CO_3.$ $10H_2O$	Sodium carbonate crystals (washing soda)
CO_2	Carbon dioxide		

6.3 The Ability of Atoms to Combine; Valency

Why does hydrogen combine with chlorine to form HCl, with oxygen to form H_2O, with nitrogen to form NH_3 and with carbon to form CH_4? The reason is that the atoms Cl, O, N and C can form different numbers of chemical bonds.

$$Cl— \qquad\qquad —O—$$

An atom of chlorine can form 1 bond: chlorine has a valency of 1.

An atom of oxygen can form 2 bonds: oxygen has a valency of 2.

$$\overset{|}{—N—} \qquad\qquad \overset{|}{\underset{|}{—C—}}$$

At atom of nitrogen can form 3 bonds: nitrogen has a valency of 3.

An atom of carbon can form 4 bonds: carbon has a valency of 4.

An atom of hydrogen can form one bond: H—: hydrogen has a valency of 1. When hydrogen combines with chlorine, oxygen, nitrogen and carbon, the compounds formed have the formulas:

$$H—Cl \qquad H—O—H \qquad H—\overset{\textstyle H}{\overset{|}{N}}—H \qquad H—\overset{\textstyle H}{\overset{|}{\underset{|}{\underset{\textstyle H}{C}}}}—H$$

These are the compounds hydrogen chloride, HCl, water, H_2O, ammonia, NH_3 and methane, CH_4.

6.4 Formulas

The formulas of the compounds HCl, H_2O, NH_3 and CH_4 are different because the valencies of the elements Cl, O, N and C are different. To work out the formula of the compound, you use the symbols and valencies of the elements in the following way.

Element:	Hydrogen	Chlorine
Symbol:	H	Cl
Valency:	1	1
Formula:		HCl

Element:	Hydrogen	Oxygen
Symbol:	H	O
Valency:	1	2
Formula:		H_2O

How did you get the formula H_2O? You *exchanged the valencies.*

Hydrogen has valency 1; oxygen has valency 2; therefore there are 2 hydrogen atoms and 1 oxygen atom in the formula, H_2O. We write O, not O_1.

Here are some more examples of this method of finding a formula.

Compound:	*Magnesium chloride*	
Element:	Magnesium	Chlorine
Symbol:	Mg	Cl
Valency:	2	1
Exchange the valencies:	1	2
Formula:		$MgCl_2$

Compound:	*Aluminium chloride*	
Element:	Aluminium	Chlorine
Symbol:	Al	Cl
Valency:	3	1
Exchange the valencies:	1	3
Formula:		$AlCl_3$

Compound:	*Calcium oxide*	
Element:	Calcium	Oxygen
Symbol:	Ca	O
Valency:	2	2
Exchange the valencies:	2	2
Formula:		CaO

(Note: You divide through by 2: CaO, not Ca_2O_2)

You can deal with groups in the same way.

<table>
<tr><td>Compound:</td><td colspan="2">Sodium sulphate</td></tr>
<tr><td>Element or group:</td><td>Sodium</td><td>Sulphate</td></tr>
<tr><td>Symbol or formula:</td><td>Na</td><td>SO_4</td></tr>
<tr><td>Valency:</td><td>1</td><td>2</td></tr>
<tr><td>Exchange the valencies:</td><td>2</td><td>1</td></tr>
<tr><td>Formula:</td><td colspan="2">Na_2SO_4</td></tr>
</table>

<table>
<tr><td>Compound:</td><td colspan="2">Iron(II) sulphate</td></tr>
<tr><td>Element or group:</td><td>Iron(II)</td><td>Sulphate</td></tr>
<tr><td>Symbol or formula:</td><td>Fe</td><td>SO_4</td></tr>
<tr><td>Valency:</td><td>2</td><td>2</td></tr>
<tr><td>Exchange the valencies:</td><td>2</td><td>2</td></tr>
<tr><td>Divide through:</td><td>1</td><td>1</td></tr>
<tr><td>Formula:</td><td colspan="2">$FeSO_4$</td></tr>
</table>

<table>
<tr><td>Compound:</td><td colspan="2">Iron(III) sulphate</td></tr>
<tr><td>Element or group:</td><td>Iron(III)</td><td>Sulphate</td></tr>
<tr><td>Symbol:</td><td>Fe</td><td>SO_4</td></tr>
<tr><td>Valency:</td><td>3</td><td>2</td></tr>
<tr><td>Exchange the valencies:</td><td>2</td><td>3</td></tr>
<tr><td>Formula:</td><td colspan="2">$Fe_2(SO_4)_3$</td></tr>
</table>

For practice, work out the formulas of these compounds. Refer to Table 6.2 for symbols and valencies.

(a) potassium chloride   (b) potassium oxide
(c) potassium sulphate   (d) calcium bromide
(e) calcium sulphate   (f) iron(II) chloride
(g) iron(III) chloride   (h) aluminium iodide
(i) aluminium oxide

6.5 Equations

In a chemical reaction, atoms are neither created nor destroyed. The atoms we start with are the same as the atoms we finish with, both in type and number. Old bonds are broken and new bonds are formed as the atoms enter into different arrangements, but the atoms themselves are unchanged. By using the symbols of the elements taking part in a reaction, we can show precisely what happens in a chemical reaction. We call this way of describing a reaction a **chemical equation**. Previously we have used word equations, such as:

$$\text{Copper} + \text{Sulphur} \longrightarrow \text{Copper sulphide}$$

Now, we shall use symbol equations:

$$Cu + S \longrightarrow CuS$$

$$\text{Sulphur} + \text{Oxygen} \longrightarrow \text{Sulphur dioxide}$$

$$S + O_2 \longrightarrow SO_2$$

$$\text{Copper} + \text{Oxygen} \longrightarrow \text{Copper oxide}$$

$$Cu + O_2 \longrightarrow CuO$$

We must use O_2 to represent a molecule, not O for an atom of oxygen because the oxygen in the air is in the form of molecules of oxygen, O_2. The two sides of an equation must be equal. Looking at the last equation to see whether the two sides are equal, we find that the left-hand side has one copper atom and two atoms of oxygen, whereas the right-hand side has one copper atom and one oxygen atom.

We must make the two sides equal. To give two oxygen atoms on the right-hand side, we need 2CuO.

$$Cu + O_2 \longrightarrow 2CuO$$

We now see that the left-hand side has one copper atom, and the right-hand side has two. We put two copper atoms on the left-hand side:

$$2Cu + O_2 \longrightarrow 2CuO$$

This is a **balanced** chemical equation. Check again.

On the left hand side, number of Cu atoms = 2; number of O atoms = 2.

On the right hand side, number of Cu atoms = 2; number of O atoms = 2.

The equation is balanced.

We can very easily put a little more information into an equation by adding letters in brackets to tell us the physical state each chemical is in: (s) means solid, (l) means liquid, (g) means gas, and (aq) means aqueous (water) solution. These letters are called **state symbols**. Putting state symbols into the equations we have written, we obtain:

$$Cu(s) + S(s) \longrightarrow CuS(s)$$

$$S(s) + O_2(g) \longrightarrow SO_2(g)$$

$$2Cu(s) + O_2(g) \longrightarrow 2CuO(s)$$

Including the state symbols tells us that the products copper sulphide and copper oxide are solids and the product sulphur dioxide is a gas.

Try writing equations for some of the reactions you have met.

(1) Zinc + Sulphur ⟶ Zinc sulphide

(2) Iron + Sulphur ⟶ Iron(II) sulphide

(3) Carbon + Oxygen ⟶ Carbon dioxide

(4) Aluminium + Iodine ⟶ Aluminium iodide

(5) Zinc + Oxygen ⟶ Zinc oxide

(6) Sodium + Oxygen ⟶ Sodium oxide

Questions on Chapter 6

1. What do you understand by these terms: (a) element, (b) compound, (c) atom, (d) molecule?

2. Give the symbols for these elements:

Zinc	Copper	Chlorine	Nitrogen
Lead	Calcium	Oxygen	Potassium
Iron	Iodine	Sodium	Magnesium

3. What do you understand by the word atom? What three things did Dalton say cannot happen to atoms? In a chemical reaction, what happens to the atoms which take part?

4. In the square below are the names of 23 elements all written across or down – for instance, Tin. Now study the square, and see if you can find the other 22.

T	I	N	C	H	L	O	R	I	N	E	H
P	N	E	O	N	E	Z	S	N	X	X	Y
H	I	R	O	N	A	I	O	D	X	S	D
O	T	G	O	L	D	N	D	I	M	I	R
S	R	X	C	X	X	C	I	U	A	L	O
P	O	T	A	S	S	I	U	M	G	V	G
H	G	Z	R	U	X	X	M	X	N	E	E
O	E	I	B	L	C	O	P	P	E	R	N
R	N	N	O	P	I	R	O	N	S	X	X
U	X	C	N	H	X	B	A	R	I	U	M
S	X	A	L	U	M	I	N	I	U	M	X
F	L	U	O	R	I	N	E	X	M	X	X

5. Give the formula for each of these compounds:

Ammonia	Sodium hydroxide
Carbon dioxide	Sodium sulphate
Calcium oxide	Copper oxide
Zinc hydroxide	Calcium hydroxide
Copper chloride	Potassium carbonate
Sodium nitrate	Copper sulphate

6. What is the number of atoms in each of these formulas?

$NaNO_3$	$AlCl_3$	$CuSO_4.5H_2O$
Na_2SO_4	$Al(OH)_3$	$Na_2CO_3.10H_2O$
$CaSO_4$	$Pb(OH)_2$	$MgSO_4.7H_2O$
$NaNO_3$	$Al_2(SO_4)_3$	$BaCl_2.2H_2O$

Crossword on Chapter 6

Trace or photocopy this grid (teacher, please see note at front of book), and fill in the answers.

Across

3 Scientists try to get these right (5)
7 A substance which cannot be split (7)
8 A way of representing a chemical reaction (8)
9 Symbol for tellurium (2)
10 A set of symbols and numbers (7)
14, 17 down, 6 down This is what Dalton said about the atoms in an element (3,3,4)
16 Atoms cannot be this (7)
17 Symbol for thulium (2)
18 This may be a hazard in the lab if long (4)
21 Symbol for helium (2)
22 Do this with equations (7)

Down

1 The smallest particle of a compound (8)
2 Same clue as 16 across (9)
3 Symbol for iron (2)
4 Nameless person (4)
5 Control technology (2)
6 See 14 across
11 A boy (3)
12 Symbol for aluminium (2)
13 Symbol for scandium (2)
15 Often used on acidic soil (4)
17 See 14 across
19 Symbol for rubidium (2)
20 Symbol for thallium (2)

7 Carbon and Carbonates

7.1 Diamond, Graphite and Charcoal

Diamond The Great Star of Africa is the world's largest cut diamond. It sparkles in the head of the Royal sceptre. Weighing 100 g, it is part of the Cullinan diamond. This 400 g stone was as big as the fist of Frederick Wells, who dug it out of a South African mine in 1905, using only his penknife. Another large diamond, Cullinan II, was cut from this great stone. It is part of the British Imperial State Crown, which the Queen wears every time she opens Parliament.

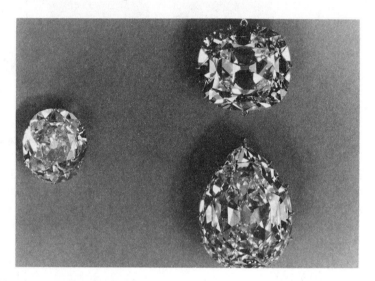

Figure 7.1   Crown jewels

It is still possible to find great gems. The 'Lesotho diamond' was found in 1967 by Ernestine Ramaboa. She and her husband worked a small diamond claim in the African country of Lesotho. The couple walked 177 kilometres (110 miles) to the capital city, Maseru, to sell the stone. For the 120 g diamond, they were paid £200 000. A diamond cutter in New York split it into 18 sparkling gems. Expert cutting brings out the sparkle and 'fire' of a diamond.

With their brilliance and 'fire', diamonds have fascinated people for 2000 years. The sparkle of diamonds is due to their ability to reflect light. The 'fire' of a diamond is due to its ability to split light into flashes of colour. Diamond is the hardest of natural materials. Because diamonds were thought to be indestructible, they were believed to keep their wearers safe from bad luck. For hundreds of years, diamonds were regarded as magical stones. Around 1800, a British chemist showed that diamond is not magical; it is a form of the

92

element carbon. When people found out that diamond is a form of carbon, there was a rush to cash in on the discovery. It turned out to be very difficult to make diamond from other forms of carbon. It was not until the 1950s that diamonds were *synthesised* (made artificially). Very high temperatures and pressures are needed. Now, 20 tonnes of synthetic diamonds are produced every year throughout the world.

Figure 7.2   Machining a piston with a diamond-tipped cutting tool

Synthetic diamonds are used to make tools. Diamond-edged tools are used for cutting glass, concrete and metal. Prospectors use diamond-tipped drills for boring through rock when they are looking for oil. Surgeons use sharp diamond-edged knives for delicate work. These knives are so sharp that they never make a jagged cut.

Figure 7.3   Diamond-studded drilling bits

Graphite

Graphite is a dark grey shiny solid. When you touch it, some of the graphite rubs off on your fingers. This is why graphite is used as a lubricant. By rubbing off on moving surfaces, it reduces the friction between them. Graphite is another form of carbon. It is unusual for a non-metallic element in that it conducts electricity, as you found out in Experiment 3.1. It is used in the manufacture of electrical equipment. Another use for graphite is to mix it with clay and manufacture pencil 'leads'. (These are so called because lead also is soft enough to mark paper.)

Diamond and graphite differ in their physical characteristics. One is a hard, brilliant gem. The other is so soft that it is used as a lubricant. Diamond is a non-conductor of electricity. Graphite is an electrical conductor. In their chemical reactions, however, diamond and graphite are the same. Crystalline forms of an element which are physically different but chemically identical are called **allotropes**. Diamond and graphite are *allotropes* of carbon. The reason why they differ is illustrated in Figure 7.4, which shows the different arrangements of carbon atoms in diamond and graphite.

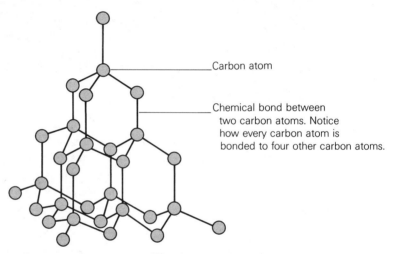

Carbon atom

Chemical bond between two carbon atoms. Notice how every carbon atom is bonded to four other carbon atoms.

(a)   The arrangement of carbon atoms in diamond

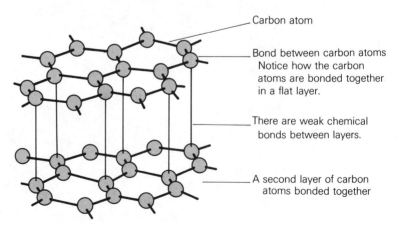

Carbon atom

Bond between carbon atoms Notice how the carbon atoms are bonded together in a flat layer.

There are weak chemical bonds between layers.

A second layer of carbon atoms bonded together

(b)   The arrangement of carbon atoms in graphite

Figure 7.4   Models of (a) diamond and (b) graphite

Charcoal   **Wood charcoal** is an impure form of carbon. It is made by burning wood in a limited supply of air. Charcoal is used by artists as a sketching material and also as a fuel for barbecues.

Animal charcoal is made by heating bones in the absence of air. It is used as an *adsorbent*. The surface of charcoal will attract many substances, which are then said to be *adsorbed* (not *ab*sorbed) on the surface. It is used for adsorbing the colour from brown sugar to give white sugar. In industry, workers sometimes have to deal with poisonous gases. They wear gas masks which contain charcoal to adsorb the gases.

Soot   Soot is an impure form of carbon.

7.2 Combustion of Carbon

Experiment 7.1

To find out what is formed when charcoal burns

1. Set up the apparatus shown in Figure 7.5.

2. Heat the charcoal and draw air through the apparatus with the suction pump. Watch for any change in the limewater.

3. As a control experiment, draw air directly into a fresh sample of limewater for the same time as you ran the experiment.

4. What effect does air have on limewater in the control experiment?

   Why do you need to do a control experiment?

   What happens to the limewater when the air bubbling through it has passed over heated charcoal?

   Describe what happens to the charcoal. What makes you think that a chemical reaction is occurring?

   What is the substance that has changed the limewater? Where has it come from?

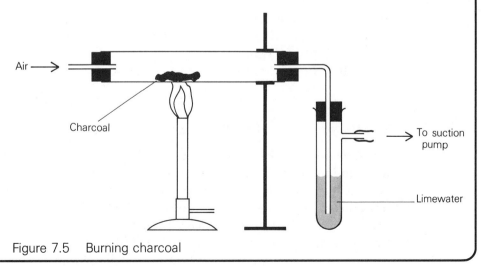

Figure 7.5   Burning charcoal

Experiment 7.2

An experiment on ourselves

1. Assemble conical flasks containing limewater as shown in Figure 7.6.

2. Six pupils line up in Row A, and six pupils line up in Row B.

3. At the word 'Go', Row A start sucking as fast as they can through the *short* tube as shown, and Row B start blowing through the *long* tube.

4. At the end of three minutes, compare the limewater in Row A with that in Row B.

5. Explain the difference.

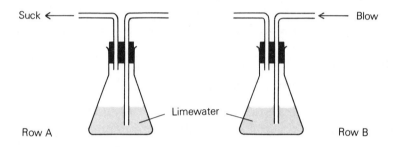

Figure 7.6   An experiment on ourselves

When you burned charcoal, you saw limewater turn 'milky'; that is to say, white particles of solid appeared in it. The change is caused by the gas which is formed when charcoal burns. Carbon combines with the oxygen in the air to form the gas carbon dioxide.

$$\text{Carbon} + \text{Oxygen} \longrightarrow \text{Carbon dioxide}$$

$$C(s) + O_2(g) \longrightarrow CO_2(g)$$

When carbon dioxide reacts with limewater (calcium hydroxide solution), the white solid calcium carbonate is formed.

$$\text{Carbon dioxide} + \text{Calcium hydroxide} \longrightarrow \text{Calcium carbonate} + \text{Water}$$

$$CO_2(g) + Ca(OH)_2(aq) \longrightarrow CaCO_3(s) + H_2O(l)$$

The test for carbon dioxide is that it turns limewater milky.

If you have done Experiment 7.2 you will have found that air sucked through limewater turns limewater slightly milky, but air which you breathe out turns limewater cloudy much faster. This shows that you must be breathing out air which contains much more carbon dioxide than ordinary air. Either you have been eating carbon or you have been eating foods which contain carbon and are oxidised to carbon dioxide in your body. You can find out by doing Experiment 7.3.

Experiment 7.3

An experiment to find out whether the foods we eat contain carbon

1. Take a piece of bread in tongs as shown in Figure 7.7 (a). Heat with a moderate flame until it has turned to a completely black mass and has stopped smoking. It is very important to heat until step (1) is complete.

2. Allow to cool, and put the black solid into an ignition tube.

3. Heat strongly until the black solid glows red. Squeeze the air out of a teat pipette, place it inside the ignition tube, and release the pressure on the teat to suck gas into the pipette. See Figure 7.7 (c).

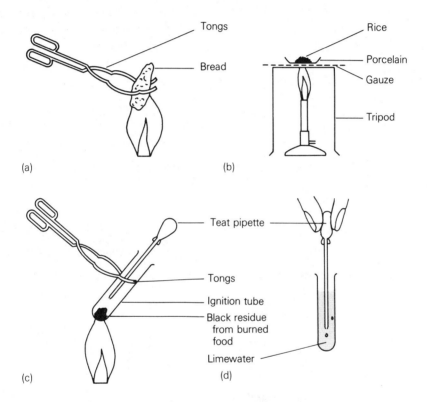

Figure 7.7   Experiment to find out whether foods contain carbon

4. Put some limewater into an ignition tube, and expel the gas from the teat pipette into the limewater, as shown in Figure 7.7 (d). Note whether it turns milky.

5. Repeat with another food. Macaroni and spaghetti can be held in tongs as in diagram (a). Rice and porridge oats can be heated on a piece of porcelain or ceramic paper as shown in Figure 7.7 (b).

6. If a food contains carbon, it will give carbon after heating by method (a) or (b), and this will be oxidised to carbon dioxide by step (c). Make a list of the foods which you have found to contain carbon.

You have found out from this experiment that various foods you eat burn to form carbon dioxide. Carbon is a fuel: when it burns, energy is given out. You fuel your body with foods such as bread and porridge oats. These are oxidised by the oxygen in the air you breathe in to give carbon dioxide, which you breathe out, and to provide you with energy. We talk about combustion of foods in the body rather than burning. **Combustion is oxidation with or without a flame; in burning there is a flame.**

All the foods, such as wheat, oats and rice, which you have experimented on are derived from plants. Plants build up these foods by the process of *photosynthesis*. *Photo* means *light* (from the Greek) and *synthesis* means *putting together*. The plant puts together carbon dioxide, taken in through the leaves, and water, taken in by the roots, and the energy of sunlight falling on the leaves. By means of the green substance in the plant called *chlorophyll*, which acts as a catalyst (see section 5.3), these three things are built up into sugar and oxygen. Sugar is converted into starch in the plant for storage. The process of photosynthesis can be represented as:

$$\text{Energy of sunlight} + \text{Carbon dioxide} + \text{Water} \xrightarrow[\text{in plants}]{\text{Chlorophyll}} \text{Sugar} + \text{Oxygen}$$

When animals eat foods containing sugar or starch, the energy stored in these foods, which came originally from the sun, is released by the process of respiration:

$$\text{Oxygen} + \text{Sugar} \longrightarrow \text{Carbon dioxide} + \text{Water} + \text{Energy}$$

Thus animals derive the energy they need from plant foods, and

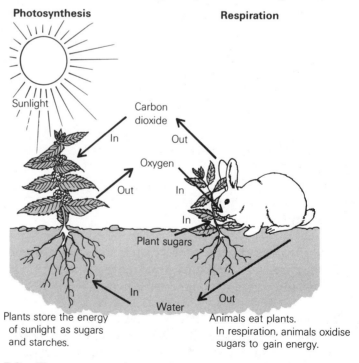

Plants store the energy of sunlight as sugars and starches.

Animals eat plants. In respiration, animals oxidise sugars to gain energy.

Figure 7.8   The energy cycle

A NEW FIRST CHEMISTRY COURSE

plants derive energy from the sun: plants **fix** the sun's energy, and animals **use** it. This energy cycle is shown in Figure 7.8.

You have seen how carbon takes oxygen from air to form carbon dioxide. In Chapter 5 you saw how metals combine with oxygen from the air. Let us now find out whether carbon has a greater or lesser liking for (in chemistry called an *affinity* for) oxygen than metals have.

7.3 Reaction of Carbon with Metal Oxides

Experiment 7.4

Will carbon take oxygen from metal oxides?

1. Assemble a number of metal oxides: zinc oxide, copper oxide, lead oxide, calcium oxide and others.

2. Take a charcoal block, and use an awl or a penknife to make a shallow hole in it.

3. Put in a little metal oxide. Dampen it with a drop of water to keep it from blowing away.

4. Light a Bunsen, and adjust it to the luminous flame; then let in a little air.

5. Taking a blowpipe, direct a flame on to the metal oxide in the carbon block. Pause for breath, and continue (see Figure 7.9)

6. If you have bushy hair, you may well singe it in this experiment. Pin it back!

7. Allow the carbon blocks to cool under the cold water tap before putting them away.

8. Tabulate your observations under the headings:

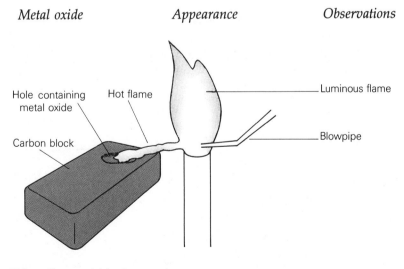

Figure 7.9   Charcoal block experiment

You will have found out from Experiment 7.4 that carbon takes oxygen away from some metals, such as copper and lead, but is unable to take oxygen away from others such as zinc and calcium:

$$\text{Carbon} + \text{Copper oxide} \longrightarrow \text{Copper} + \text{Carbon dioxide}$$

$$\text{Carbon} + \text{Lead oxide} \longrightarrow \text{Lead} + \text{Carbon dioxide}$$

The process of taking oxygen away is called **reduction**; it is the opposite of giving oxygen, which is called **oxidation**. Carbon reduces lead oxide to lead. Lead oxide oxidises carbon to carbon dioxide. The processes of oxidation and reduction always occur together. **A substance which gives oxygen is called an oxidising agent; a substance which takes oxygen is called a reducing agent.** Lead oxide acts as an oxidising agent, carbon acts as a reducing agent; in the reaction the oxidising agent is reduced, and the reducing agent is oxidised.

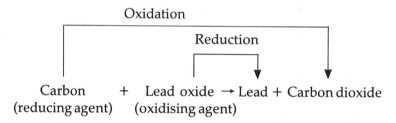

Some elements are better reducing agents than others. Carbon will reduce the oxides of metals which are less powerful reducing agents than carbon.

7.4 From Carbonates to Concrete

A carbonate is a compound which contains a metal and carbon and oxygen. Whenever you see a name ending in *-ate*, it means that the compound contains oxygen. Calcium carbonate is the most widespread of metal carbonates. It is found in three forms, marble, limestone and chalk. They look different, but are chemically the same.

Experiment 7.5

Action of heat on calcium carbonate

1. Assemble the apparatus as shown in Figure 7.10. Put the piece of marble on the edge of the gauze, where it can be heated strongly. Balance it carefully: you do not want it to fall off when hot.

2. Heat the marble with a roaring Bunsen flame for fifteen minutes. From time to time, hold a drop of limewater on the end of a glass rod over the marble. Observe the limewater.

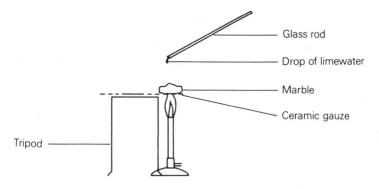

Figure 7.10   Heating calcium carbonate

3. Allow to cool. When cool, put the piece of marble into an evaporating basin. Add one drop of universal indicator. Compare the colour with a drop of universal indicator on an unheated piece of marble. What does the colour tell you?

4. Carefully, from a teat pipette, add water drop by drop. You should notice two things: (a) a gas, (b) a change on the surface of the marble. What makes you think that there is a change in energy occurring?

5. Dissolve the crumbly solid formed on the surface of the marble in water. Filter. Test the solution.

    (a) To a portion, add universal indicator.

    (b) Take a second portion, and blow into it through a straw. Do you recognise this solution?

6. Explain your observations. What gas was evolved? What formed at the surface of the marble? What happened when you added water? What solution was formed?

When calcium carbonate is heated strongly, as in Experiment 7.5, it splits up. Carbon dioxide is given off, and a white solid, calcium oxide, is formed. When you add an indicator to calcium oxide, you see it turn to its alkaline colour. When you add water to calcium oxide, you see the water turn into steam. The reason is that calcium oxide reacts with water in an **exothermic** reaction – a reaction which gives out heat. The product of the reaction is calcium hydroxide.

Calcium oxide + Water $\longrightarrow$ Calcium hydroxide

$$CaO(s) + H_2O(l) \longrightarrow Ca(OH)_2(s)$$

Calcium oxide is often called **quicklime**. The word *quick* means *living*. Early chemists thought that calcium oxide looked as though it was alive when it was reacting with water. Calcium hydroxide is often called **slaked lime**. The early chemists thought that calcium hydroxide was quicklime which had *slaked*, that is *satisfied* its thirst. A solution of calcium hydroxide is called **limewater**.

Calcium $\xrightarrow{\text{Water}}$ Calcium $\xrightarrow{\text{Water}}$ Calcium
oxide   hydroxide   hydroxide solution
(Quicklime)   (Slaked lime)   (Limewater)
$CaO(s)$ $Ca(OH)_2(s)$ $Ca(OH)_2(aq)$

Manufacture of Quicklime

Quicklime (calcium oxide) is important both in agriculture and in the building industry. It is made by heating limestone (calcium carbonate) at 1000 °C in a furnace called a **lime kiln** (see Figure 7.11). Some of the quicklime is converted into slaked lime (calcium hydroxide). Slaked lime is used in agriculture. In farms which have acidic soil, slaked lime is spread on the fields to reduce acidity and make the soil more fertile.

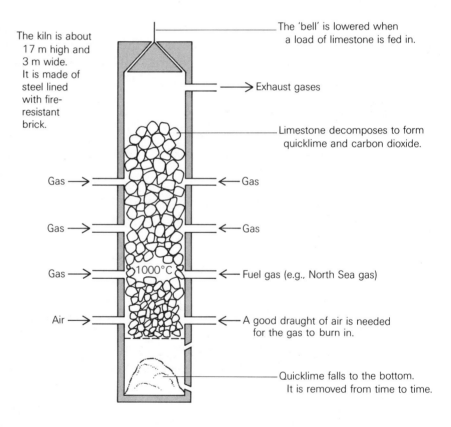

Figure 7.11   One type of lime kiln (Others use coke or oil for heating.)

- Limestone decomposes to form calcium oxide and carbon dioxide. Calcium oxide is a basic oxide. Carbon dioxide is an acidic oxide. What would happen if the carbon dioxide produced were not removed from the kiln by the through draught of air?

Manufacture of Cement

Limestone is used in the manufacture of cement. Figure 7.12 shows a view of the Blue Circle Group Works at Hope, where cement is made. Limestone is quarried in the hills behind the works. At the extreme right of the works is the very tall chimney stack. This is the preheater building. The two massive horizontal pipes which connect the preheater building with the rest of the works are the kilns. Limestone and clay are fed into the kilns, which are heated by coal or oil. As the mixture passes through the hot, rotating kilns, it is converted into cement. Figure 7.13 shows a close-up of the preheater tower and one of the kilns. In some regions, chalk is used instead of clay.

Figure 7.12   Blue Circle Group Works at Hope

Figure 7.13   Preheater tower and kiln in cement works

Concrete   Cement is used to make concrete; see Figure 7.14.

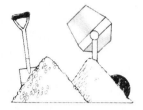  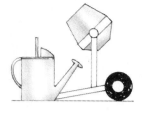

1 Sand and gravel are mixed.

2 Cement powder is added. The mixture is turned over and over in a cement mixer.

3 Water is added. Crystals start to grow in the cement. They bind the mixture together.

Figure 7.14   Making concrete

Mortar is made by mixing sand and cement. It is used for holding bricks together. Mortar gradually hardens on exposure to air. Carbon dioxide in the air converts the calcium hydroxide in mortar into calcium carbonate.

The foundations of your house are of concrete. Under the floor there is concrete to seal the house from the ground. If the house is built of bricks, the bricks are joined together by cement mortar. It may be made of concrete bricks or blocks, again joined with cement mortar. The lintels over the doors and windows are concrete beams. Concrete blocks are often used for inner walls. As well as being easy to lay, they prevent the escape of heat. Most houses are roofed with concrete tiles. You may have a concrete patio, a concrete wall, a concrete path or drive. Many garages are built from concrete slabs.

Experiment 7.6

How strong is concrete?

Here are two recipes for concrete:

Ingredients	Concrete A	Concrete B
Cement	1 bucket	1 bucket
Sand	$1\frac{1}{2}$ buckets	2 buckets
Gravel	$2\frac{1}{2}$ buckets	3 buckets
Water	$\frac{1}{2}$ bucket	$\frac{1}{2}$ bucket

Which is the stronger, Concrete A or Concrete B?

What can you do to find out? You will have to think up some way of casting a bar or disc of concrete. Then, when it has set, you will have to think up a way of testing its strength. When you have made a plan, show it to your teacher. If he or she approves, start work.

You may like to extend your work by trying other recipes for concrete.

Some Concrete Questions

1. (a) What are the chief materials used in the making of cement?

   (b) Why is there a big cement factory at Hope in Derbyshire?

   (c) What are the advantages and disadvantages in having a cement factory in your neighbourhood?

   (d) Some kilns are fuelled by North Sea gas; others are coal-fired. Why do you think the Derbyshire kilns use coal?

   (e) Limestone and chalk have the same chemical name. What is it?

   (f) What gas is formed in the cement kilns?

2. Make a list of examples of the use of concrete as a building material.

3. Mortar can be made by slaking quicklime and then mixing it with sand.

   (a) What is *quicklime*?

   (b) What is meant by *slaking*?

   (c) A chemical reaction takes place as mortar hardens. What is this reaction?

7.5 Carbon Dioxide

You can prepare carbon dioxide by the action of dilute hydrochloric acid on marble chips.

Calcium carbonate + Hydrochloric acid $\longrightarrow$ Calcium chloride + Carbon dioxide + water

$$CaCO_3(s) + 2HCl(aq) \longrightarrow CaCl_2(aq) + CO_2(g) + H_2O(l)$$

Experiment 7.7

To prepare carbon dioxide and study its properties

1. Set up the apparatus as shown in Figure 7.15.

2. Add dilute hydrochloric acid to the marble chips. Discard the first boiling tube of gas collected as it will contain a lot of air.

3. Collect two boiling tubes of gas over water as shown in method (a). Cork the tubes before taking them out of the trough. Stand them in a rack.

4. Attach a right-angled delivery tube as in method (b). Collect two gas jars full of gas by downward delivery. A slow count of fifteen should tell you when all the air has been driven out. Cover the gas jars with lids.

5. Leave the right-angled tube attached, and bubble the gas through limewater in a boiling tube.

6. Bubble the gas through a boiling tube one third full of water to which universal indicator has been added.

7. Into a boiling tube of gas, lower a burning taper.

8. Figure 7.16 (a) shows how to find out whether carbon dioxide is denser or less dense than air. Put a jar of carbon dioxide over one of air. Remove the lids, count twenty seconds, and replace the lids. Test both jars with a lighted splint. Where is the carbon dioxide?

9. Repeat (8) with carbon dioxide at the bottom and air at the top, as in Figure 7.16 (b). What happens to the carbon dioxide?

   This test can be done with stoppered boiling tubes if there are not enough gas jars for the class.

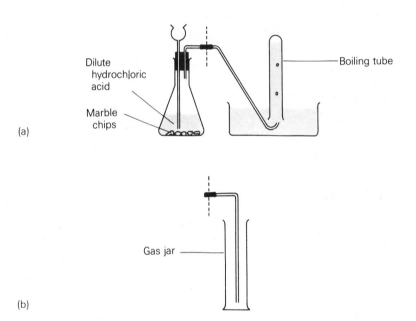

Figure 7.15   Preparation and collection of carbon dioxide

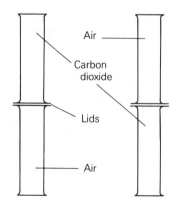

Figure 7.16   Testing the density of carbon dioxide

You have found out about carbon dioxide:

- It is a colourless, odourless gas.
- It is not very soluble in water: it can be collected over water.
- It is weakly acidic: it turns universal indicator cherry red.
- It turns limewater 'milky'. This reaction is a test for carbon dioxide.
- It extinguishes a burning taper.
- Carbon dioxide is denser than air. Test (8) shows that carbon dioxide sinks into the bottom gas jar. Test (9) shows that, when placed at the bottom, carbon dioxide will stay in the bottom jar.

Experiment 7.8

A problem with rocks

Many rocks contain calcium carbonate without being 100 per cent calcium carbonate. Examples are limestone, marble, sandstone, calcite and chalk. A challenging problem is to compare the percentages of calcium carbonate in the different rocks. Can you tackle it?

First, you have to scour the countryside for samples of rock. Then you have to work out a plan. What chemical reaction will all the kinds of rock that contain calcium carbonate undergo? How can you use this reaction to find the mass of calcium carbonate in a weighed sample of rock? Write down the measurements you will have to make. Show your plan to your teacher before you start your experiment.

Extension Problem

Is it true that some scouring powders and tooth powders contain chalk?

Is blackboard chalk really chalk?

Do eggshells contain calcium carbonate?

How can you find out?

If you find that one of these substances does contain chalk, how can you find the percentage of chalk in that substance?

Again, check with your teacher before you do your experiments.

7.6 The Carbon Cycle

Carbon dioxide makes up 0.03 per cent of the air, a very small but essential percentage. Plants take carbon dioxide from the air to use in photosynthesis. The combustion of coal and oil produces carbon dioxide. Decay of plant and animal material produces carbon dioxide. We and other animals produce carbon dioxide in respiration. Carbon dioxide dissolves in the sea and is built into the shells of sea creatures as calcium carbonate. When these creatures die, they sink to the sea bed. There they accumulate to form deposits of limestone. The balance between reactions using carbon dioxide and reactions producing carbon dioxide is called the **carbon cycle**. It is illustrated in Figure 7.17.

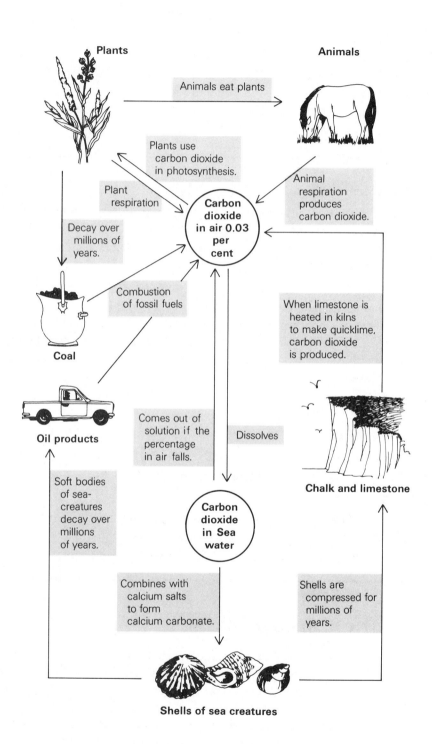

Figure 7.17   The carbon cycle

7.7 Methods of Fighting Fires

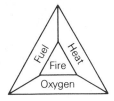

Figure 7.18   The fire triangle

The fire triangle shows the three things a fire needs: fuel to burn, heat to raise the fuel to the temperature at which it is hot enough to burn (called its *ignition temperature*), and oxygen. If any one of these sides of the triangle is removed, the fire will go out.

Removal of Fuel

A fire will burn itself out when it runs out of fuel. This technique is used in fighting forest fires. If a wide swathe of trees and bushes in the path of the fire is removed, the fire will not be able to spread and will die out.

Removal of Heat

Throwing water on to a fire cools it down. If it cools down below the ignition temperature of the fuel, the fire will go out. There are some chemical fires on which water cannot be used, for example sodium fires. Water reacts with sodium to give the flammable gas hydrogen, and this makes the fire worse. It is part of a fireman's training to learn which chemicals react with water.

Electrical fires cannot be extinguished with water. If you direct a jet of water on to a smouldering piece of electrical equipment, a current of electricity can pass along the jet of water and give you a severe shock. The method to use here is to switch off at the mains and use one of the carbon dioxide fire extinguishers.

Oil fires cannot be extinguished with water. This is why it is so dangerous if the oil tanks on a ship catch fire. The burning oil floats on top of the water, and the blaze spreads over a wide area. If the crew try to abandon the ship, they face the danger of being burned in their lifeboats.

Experiment 7.9 shows what happens when you try to extinguish an oil fire with water, and also how you can put out the fire with a damp cloth. The fire goes out because the cloth is excluding air. The cloth is dampened to prevent it from catching fire itself. This method is the one to use if a pan of cooking fat catches fire at home: switch off the gas or electricity and cover the pan with a damp towel.

For many fires, however, water is a good extinguisher. The soda–acid type of extinguisher provides a means of obtaining a powerful jet of water quickly. This type of extinguisher is the large kind which stands on the floor. Figure 7.19 shows how it works.

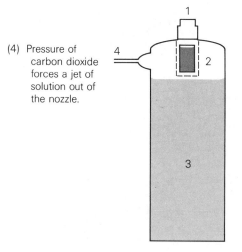

(1) Strike the knob.

(2) The knob smashes a small bottle of sulphuric acid into a metal gauze.

(3) Sulphuric acid from the broken bottle meets a solution of sodium hydrogencarbonate. They react to form carbon dioxide.

(4) Pressure of carbon dioxide forces a jet of solution out of the nozzle.

Figure 7.19   A soda–acid extinguisher

Experiment 7.9

Will water extinguish oil fires?

1. Put an evaporating basin on to a tile on the bench. Put into it 20 cm^3 of oil (cooking oil or turpentine) and, using a piece of rag as a wick, set fire to it.

2. Now direct a jet of water from a squeeze bottle on to the flames.

3. Do you find that the oil floats on top of the water and continues to burn?

4. Take a bench cloth, wet it under the tap, wring it out and put it over the evaporating basin.

Experiment 7.10

To try out a model of a soda–acid fire extinguisher

1. Fill a conical flask two thirds full of saturated sodium hydrogencarbonate solution. Fit a delivery tube through a rubber bung to reach down to near the bottom of the flask (see Figure 7.20).

2. Tie a piece of cotton thread round the mouth of an ignition tube. **Wearing safety glasses**, fill the ignition tube with concentrated hydrochloric acid, and lower it into the conical flask. Keep the acid above the level of the solution so that they do not mix. Replace the bung.

3. Now you can try out your model. Take a small piece of cotton wool in an evaporating basin. Add 10 cm^3 of methylated spirits and set fire to it.

4. Now shake the model fire extinguisher so that the acid mixes with and then reacts with the solution of sodium hydrogencarbonate. Direct the jet of water on to the fire.

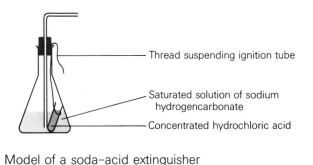

Thread suspending ignition tube

Saturated solution of sodium hydrogencarbonate

Concentrated hydrochloric acid

Figure 7.20   Model of a soda–acid extinguisher

Removal of Oxygen

The third side of the triangle is oxygen. Carbon dioxide extinguishers act by excluding oxygen. Carbon dioxide is a dense gas which does not support combustion. It forms a blanket over the fire and keeps out oxygen. Carbon dioxide is a good extinguisher for many types of fire, including oil fires and electrical fires. A fire of burning magnesium, however, would continue to burn in the gas. There are three types of extinguisher using carbon dioxide. These are *powder, gas* and *foam extinguishers.*

Figure 7.21   Three types of fire extinguisher

Powder extinguishers contain substances like sodium hydrogen-carbonate (baking soda), which decomposes in the heat of the fire to give carbon dioxide. This drives away the air, and the fire goes out. Powder extinguishers are useful for electrical fires. They will not cope with a big fire. They are small in size and usually placed in a bracket on the wall. Experiment 7.11 uses a powder extinguisher.

Cylinders of carbon dioxide stored under pressure are useful fire extinguishers. Figure 7.22 shows an extinguisher of this type. It is usually operated by removing a pin and squeezing the trigger. A jet of carbon dioxide gas comes out of the nozzle for about two minutes. You must never use a fire extinguisher except when it is needed for putting out a fire. If there is a cylinder of carbon dioxide in your store, your teacher may demonstrate how it works on some burning kerosene.

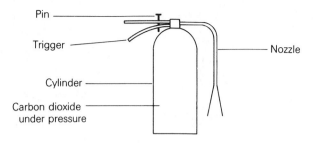

Figure 7.22   A carbon dioxide fire extinguisher

Experiment 7.11

To use a model powder extinguisher

1. Get a little kerosene (called 'paraffin' by retailers) burning, with a wick, in an evaporating basin.

2. With a spatula, sprinkle sodium hydrogencarbonate on to it.

3. Is the fire extinguished?

Experiment 7.12

To use a model foam extinguisher

1. Assemble the model shown in Figure 7.23. The bottle contains 100 cm^3 saturated sodium hydrogencarbonate solution containing 1 g saponin, a foam stabiliser. The test tube contains 10 cm^3 of dilute hydrochloric acid.

2. Start a small fire in an evaporating basin, using 10 cm^3 kerosene (paraffin) and a wick (a piece of rag or filter paper).

3. Turn the model foam extinguisher upside down, and direct the stream of foam on to the fire.

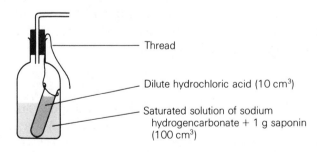

Figure 7.23   A model foam extinguisher

Foam extinguishers employ carbon dioxide. Carbon dioxide bubbles mixed with foam form a heavier layer on top of the fire than carbon dioxide alone. This type of extinguisher is large and usually stands on the floor. When needed, it is turned upside down and tapped sharply on the floor to push in a knob. The knob releases an acid solution into a solution of sodium hydrogencarbonate and *saponin*, a foam stabiliser. A jet of foam consisting of the solution and carbon dioxide shoots out. Experiment 7.12 shows you how to make and use a model foam extinguisher.

A foam extinguisher is unsuitable for those types of chemical fires which react with water.

Experiment 7.13

To use carbon dioxide gas as a fire extinguisher

1. Assemble the equipment shown in Figure 7.24. Have to hand a bottle of dilute hydrochloric acid.

2. Set fire to the kerosene.

3. Pour acid carefully down the side of the big beaker on to the marble chips.

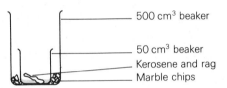

- 500 cm³ beaker
- 50 cm³ beaker
- Kerosene and rag
- Marble chips

Figure 7.24   Using carbon dioxide on a fire

There are gases other than carbon dioxide which do not support combustion and which are denser than air. BCD extinguishers contain bromochlorodifluoromethane. On hot surfaces, this liquid produces a dense vapour which excludes air from the fire. BCD is used because it is chemically unreactive, even at high temperature.

7.8 Fuels

Coal   Many of the fuels we use contain carbon and carbon compounds. Coal is one. Most of the coal which is mined is used in power stations.

Figure 7.25   Coal-cutting in a modern mine

Petroleum   Petroleum oil is a mixture of a large number of compounds. Most of them are hydrocarbons (compounds of carbon and hydrogen). Crude oil is separated into a number of different products by fractional distillation. Figure 2.10 shows the fractions that are obtained.

Petroleum gases are sold as bottled gases, such as Camping Gaz. There is a huge demand for the liquid fuels: petrol for motor vehicles, kerosene for jet engines and rockets, aeroplane fuel, gas oil for heavy vehicles like tractors and diesel oil for trains.

Figure 7.26   An oil tanker

With so many hydrocarbon fuels being burnt, it is important to know what is formed when hydrocarbons burn. Experiment 7.14 uses a candle made from paraffin wax. This is one of the products obtained from the fractional distillation of crude oil.

Experiment 7.14

To find out what is formed when a candle burns

1. Set up the apparatus shown in Figure 7.27. The suction pump draws a stream of air through the apparatus. It carries any gases produced by the burning candle through the cooled U tube and through the limewater.

2. Watch for any change in the white anhydrous copper sulphate in the U tube and in the limewater.

3. Do a control experiment without the candle for the same length of time.

4. What change have you seen in the limewater? What substance has caused this change? Which element is oxidised to the substance you have detected?

5. What change have you seen in the anhydrous copper sulphate? What substance caused this change? Which element is oxidised to this substance?

6. Which two elements have you detected in candle wax? Is wax a mixture or a compound of these elements?

7. What does the control experiment tell you?

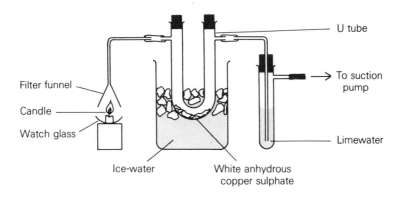

Figure 7.27   Testing products from a burning candle

All petroleum fractions are mixtures of hydrocarbons, compounds of carbon and hydrogen. They all burn to form carbon dioxide and water. If the supply of air is limited when hydrocarbons burn, combustion may not be complete. Instead of burning to form carbon dioxide, CO_2, they may burn to form carbon monoxide, CO, and soot, which is a form of carbon. You will have noticed some soot in the candle experiment. Carbon monoxide is a poisonous gas. Running a car engine in a closed garage does not provide enough air for the fuel to burn completely to carbon dioxide and water. Instead, it burns incompletely, and the exhaust fumes will contain carbon monoxide. Soon there will be a dangerous level of this poisonous gas.

If fuel oils contain sulphur compounds, the poisonous gas sulphur dioxide will be formed when the oils are burned. The oil burned in domestic heaters is called *kerosene* in the oil industry and *paraffin* by the retailers. It must be carefully freed from sulphur compounds. The fuel burned by power stations is allowed to have a higher sulphur content. Every power station sends tonnes of sulphur dioxide into the air every day. To remedy the situation, power stations will have to install anti-pollution devices. You read about these on pp. 70–71. The industry can do this if people are prepared to pay more for electricity.

Natural Gas

Natural gas is the name given to the gas which always occurs together with crude oil. Since it was discovered in the North Sea, natural gas has been piped into Britain and used for cooking and heating. Natural gas is mainly methane, a compound of formula CH_4.

The Future

Coal and oil and natural gas are **fossil fuels**. They are the fossilised remains of living things. Coal was formed 240 million years ago. The Earth was covered with swamps and forests. Bacteria acted on dead and decaying trees and plants, and turned them into peat. The peat became covered with layers of mud, and decayed slowly in the absence of air. Under the pressure of the layers on top of it, the peat slowly turned into coal. The decay took millions of years. When we have used up the Earth's store of coal, no more will be produced.

Oil was formed from sea-living creatures. When they died, their remains became covered with sand and rock. They decayed at high temperature and high pressure over millions of years. Natural gas was formed in a similar way.

People reckon that supplies of North Sea oil and gas will be used up in about 20 years' time. In other parts of the world, there is a similar forecast. The UK has enough coal to last for 200 years. Many other countries still have large deposits of coal. In the future, coal may become an even more important fuel. Scientists have found methods of turning coal into gaseous fuels. These gaseous fuels can be burnt in central heating systems and industrial furnaces. They could replace natural gas. There are also methods of turning coal into liquid fuels. South Africa has no oil and has cheap coal. In South Africa, there is a big plant which makes liquid fuels from coal. The liquid fuels can be used in vehicle engines and in industrial furnaces. They can replace petrol and kerosene.

Using coal to replace oil and gas will help us in the twenty-first century, but the Earth's coal will not last for ever. The shortage of fuels which is foreseen is called **the energy crisis**. Scientists are working on the problem. Many countries are experimenting with methods of generating electricity by using windmills. Windpower is a *renewable* source of energy. We cannot use it up. Another renewable source of energy are the tides. The tides can be used to drive turbines

Figure 7.28   Drilling for oil

Figure 7.29   A Shell oil refinery

and generate electricity. Countries which have waterfalls use them to generate electricity. Many countries are investing in **nuclear power**. For a fuller treatment of the energy crisis, see *Extending Science 7: Energy* by J. J. Wellington (ST(P)) and *Extending Science 9: Nuclear Power* by R. E. Lee (ST(P)).

Which Fuel?

1. (a) Name three fossil fuels. Briefly explain how they were formed.

   (b) Explain why fossil fuels are important to us.

   (c) What do people mean by *the energy crisis*? What plans do people have for coping with the energy crisis?

   (d) What is meant by a *renewable* energy source?

2. (a) Name two countries which are producers of petroleum oil.

   (b) What kind of compounds does petroleum oil contain?

   (c) How is crude oil separated into useful substances? (Revise Chapter 2 if necessary.) Give the names and uses of four substances obtained from crude oil.

   (d) What is formed when these substances burn (i) in plenty of air (ii) in a poor air supply?

   (e) Why must you be careful to open a window if you use a gas-burning water heater in the bathroom?

Questions on Chapter 7

1. Write down a word or words to fill each blank:

   The hardest naturally occurring substance is _____. It is a form of the element _____. It is used in jewellery because _____, and it is used for industrial purposes such as _____. The other pure form of this element is _____. When baked with clay, this is used to make _____. It conducts electricity, and is therefore used for _____. Since it is soft and flaky, it is used as _____.

2. Write down a word or words which will correctly fill the blanks in this passage:

   Some foods that we eat contain starch. Examples are _____, _____ and _____. When these foods are burned in the laboratory a black mass of _____ is left. This can be oxidised on heating to form the gas _____, which turns limewater _____. In our bodies, the starchy foods we eat are oxidised by _____ in the air we breathe in. This is why we breathe out _____. The energy stored in starch is put there by the process of _____, in which plants use _____ and water, and the energy of sunlight. This process is catalysed (helped) by _____ in the plants.

3. A baker buys flour from a supplier. He suspects that the supplier has added powdered chalk to the flour, although this practice is illegal. How can the baker test to see whether his suspicions are correct?

What will happen to the chalk when the bread is baked?

4. On 12 February, 1987, newspapers carried a report of an inquest on a woman who suffocated after falling asleep in front of her gas fire. Workmen later removed three buckets full of birds' nesting materials from the top of the chimney.

Can you explain what caused the woman's death?

5. Mortar consists of a wet mixture of calcium hydroxide, sand, and animal hair. If some old mortar is powdered and treated with dilute acid, carbon dioxide is formed. Where do you think the carbon dioxide *originally* came from?

6. (a) What are the two liquids in a soda–acid extinguisher? Which gas do they react to form?

(b) How does water extinguish a fire? Explain why water cannot be used to extinguish a pan of burning oil.

(c) How would you extinguish a chip pan fire if no extinguisher were available?

(d) Name two types of extinguisher which are safe to use on burning oil.

(e) Give an example of another kind of fire on which water cannot be used. Explain why.

(f) Make a list of ten simple precautions that can be taken in the home to prevent fire.

7. Write down the missing words:

Adding oxygen to a substance is called _____.

Taking oxygen from a substance is called _____.

When carbon reacts with copper oxide to form copper, carbon is acting as a _____ agent. Carbon is _____ to carbon dioxide, and copper oxide is _____ to copper.

8. You are supplied with gas jars containing carbon dioxide. Describe experiments you can do to show that:

(a) carbon dioxide is denser than air;

(b) carbon dioxide dissolves in sodium hydroxide solution;

(c) carbon dioxide will not support combustion.

9. Name eight substances which are obtained from petroleum oil, and give their uses.

10. What are the two chief substances produced when petrol burns?

What other substance is produced if the supply of air is restricted?

What pollutant is formed in small amounts when petrol burns?

11. Explain the meanings of the following:

(a) oxidation   (b) reduction   (c) fossil fuel   (d) natural gas
(e) energy crisis   (f) crude oil   (g) petroleum gas   (h) foam
extinguisher   (i) concrete   (j) lime kiln   (k) photosynthesis
(l) allotropy.

12. Trace or photocopy the grid (teacher, please see note at the front of
the book). Write the answer to each clue across the row opposite
each number.

Clues

1 This is the source of petrol in
Britain (5,3,3)

2 A carbon dioxide _____ is handy
in case of fire (12)

3 This fuel is used in trains and heavy
lorries (6,3)

4 People worry that when the oil runs
out there will be an _____
_____ (6,6)

5 This petroleum product will give
you a smooth-running engine
(11,3)

6 A pollutant from power stations
which burn coal or oil (7,7)

7 All petroleum fractions contain these
compounds (12)

8 A pollutant from vehicle engines
(6,8)

9 If the oil runs out, we can still get
electricity from this (7,5)

10 These harness a renewable energy
source (9)

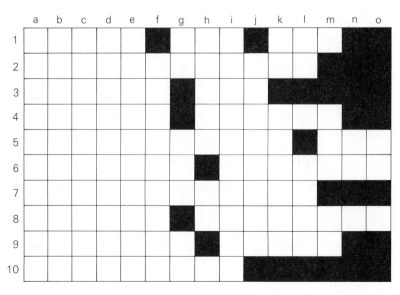

Use these letters from the words you have filled in to find out what the motorist likes to do.

6d	6b	1d

2i	3h	10e	2a

9i	3c	5h	4i	7e	9d

10b	9a

2c	7a	8o

4h	1i	8c

Crossword on Chapter 7

Trace or photocopy this grid (teacher, please see note at the front of the book) and then fill in the answers.

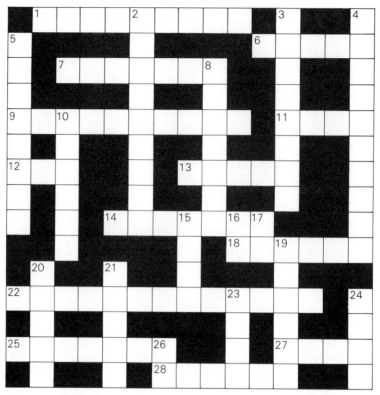

Across

1 One form of carbon is used in this (9)
6 Tool sometimes tipped with 7 across (5)
7 The hardest naturally occurring material (7)
9 A process which resembles burning, without a flame (10)
11 Given out in 9 across (4)
12 Listen with this (3)
13 It is given out when magnesium burns (5)
14 7 across will _____ 13 across (7)
18 Sweet substances containing carbon (6)
22 These burn to give carbon dioxide and 20 down (13)
25 Carbon _____ copper oxide to copper (7)
27 Sodium hydroxide is caustic _____ (4)
28 You obtain this from foods containing 22 across (6)

Down

2 A name for calcium carbonate (9)
3 Slippery form of carbon (8)
4 3 down and 7 across are a pair of these (10)
5 We need limestone to make this building material (8)
8 Carbon _____ is formed when carbon burns (7)
10 Beautiful form of calcium carbonate (6)
15 Name given to a mixture of clay and 3 down (4)
16 Symbol for caesium (2)
17 The French for *you* (2)
19 3 down feels like this (6)
20 Formed by 9 across of 22 across (5)
21 It gives you 13 across (and some 11 across) (5)
23 A jelly-like substance, used in biology laboratories (4)
24 This fuel is burned in moorland areas of Ireland (4)
26 Symbol for selenium (2)

8 Water

8.1 The Water Cycle

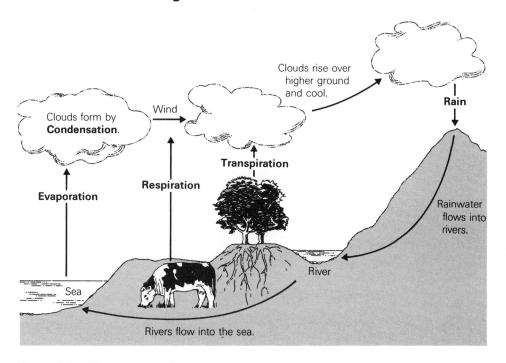

Figure 8.1   The water cycle

Figure 8.1 shows why we never run out of water. When rain falls, it trickles through the ground to drain into rivers and lakes. Rivers eventually flow into the sea. There are three ways in which the supply of water in the atmosphere is replenished. Water vapour passes up into the atmosphere from below by evaporation from lakes, rivers and seas. The process of transpiration in plants draws water from the soil through the roots and allows water to evaporate through the leaves. In respiration, animals burn up starchy food materials in their bodies and breathe out air containing much water vapour. Thus, respiration, transpiration and evaporation send water vapour into the upper atmosphere. When the air becomes saturated with water vapour, rain clouds form. When the clouds cool, water vapour condenses and falls to earth as rain.

Types of Natural Water

Rain water. Rain water is the purest naturally occurring form of water. As it falls through the air, it dissolves the gases oxygen, nitrogen and carbon dioxide. Natural rain is therefore a weakly acidic solution. You read about *acid rain* in Chapter 5.

Ground water and spring water. Rain water trickles through the soil until it meets a layer of rock which it cannot penetrate. There it accumulates as **ground water**. Ground water is fairly pure water. Many Water Authorities pump it to the surface and use it after a small amount of purification. If the ground water finds a crack in the rock above it, it may surface as a spring. **Spring water** is usually drinkable. As it trickles along underground rocks, spring water may dissolve minerals from the rocks. A spring at Epsom was found to contain magnesium sulphate. Some people said they felt much better after drinking Epsom spring water. Actually, magnesium sulphate is a laxative, and has been known ever since as 'Epsom salts'.

River water and lake water. As rain water flows towards rivers and lakes, it dissolves salts from the rocks it meets. If it trickles through rocks containing calcium and magnesium sulphates, these salts dissolve in it. Calcium and magnesium chlorides are also likely to be present in river water and lake water. Chalk and limestone (calcium carbonate) are insoluble in water. When water which contains dissolved carbon dioxide trickles through chalk or limestone, a chemical reaction happens. The soluble salt calcium hydrogen-carbonate is formed.

$$\text{Calcium carbonate} + \text{Water} + \text{Carbon dioxide} \longrightarrow \text{Calcium hydrogencarbonate}$$

$$CaCO_3(s) + H_2O(l) + CO_2(g) \longrightarrow Ca(HCO_3)_2(aq)$$

A similar reaction happens with magnesium carbonate.

Sea water. Sea water contains up to 4% of dissolved solids. Present are sodium chloride, magnesium chloride, magnesium sulphate, potassium bromide and other salts.

8.2 We Use Water

We each use about 160 litres of water a day. Figure 8.2 shows how this total is made up.

Washing and baths

50 litres

Lavatory

50 litres

Waste: dripping taps, leaking pipes

20 litres

A NEW FIRST CHEMISTRY COURSE

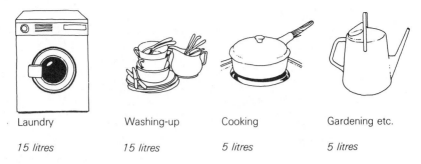

Laundry	Washing-up	Cooking	Gardening etc.
15 litres	*15 litres*	*5 litres*	*5 litres*

Figure 8.2   We use water

This is not the end of our needs. Farmers use water to grow our food. Industry uses water to make clothing, books, cars and all our other possessions. Power stations use water in generating electricity. Table 8.1 shows where the water goes.

Table 8.1   *Water needed to make 1 tonne of product*

Product	*Tonnes of water needed*
Bread	4
Coal	5
Paper	90
Steel	45
Sugar	8
Nylon	140

The total quantity of water we use adds up to about 80 000 litres a year per person. Much of the water is recycled. Chemical manufacturers and power stations use a lot of water for cooling. They take water from a stream or river, use it and return it to the source as warm water.

8.3 Water Treatment Works

The water we use comes from rivers and lakes. Often it is stored in reservoirs. It is treated in a water works to make it safe to use. Figure 8.3 shows what is done.

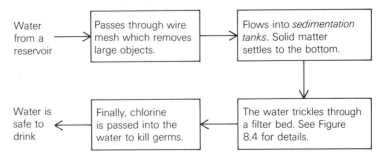

Figure 8.3   The treatment which water receives at a water works

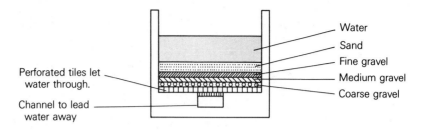

Figure 8.4   A water filter bed

You can make a model water filter bed from a plastic juice bottle. Figure 8.5 shows how.

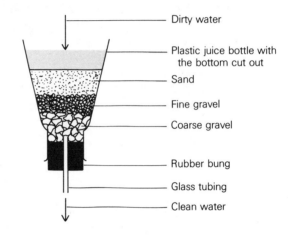

Figure 8.5   A model water filter bed

8.4 Sewage Works

After use, water passes into sewers and flows to the sewage works. There it is made pure enough to be discharged into a river or the sea.

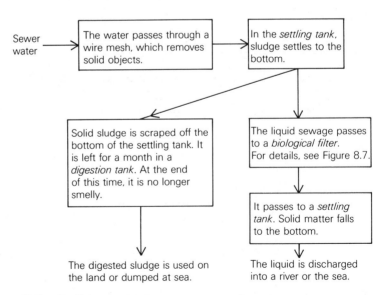

Figure 8.6   Purifying water in a sewage works

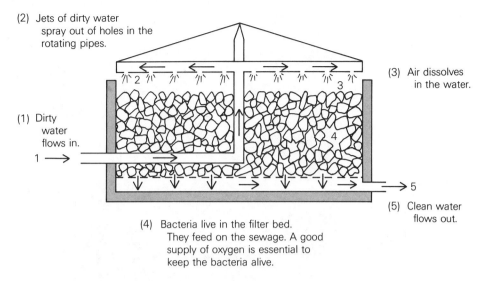

(2) Jets of dirty water spray out of holes in the rotating pipes.

(3) Air dissolves in the water.

(1) Dirty water flows in.

(4) Bacteria live in the filter bed. They feed on the sewage. A good supply of oxygen is essential to keep the bacteria alive.

(5) Clean water flows out.

Figure 8.7   A biological filter bed

When water is discharged from the sewage works into a river, it is not pure enough to drink. Bacteria in the river purify the water further. They feed on solid matter and germs. The water must contain enough dissolved oxygen to keep the bacteria alive.

Sometimes, untreated sewage is discharged into rivers and into the sea. This is because the population keeps increasing. The sewage works cannot cope with all the sewage produced and all the industrial waste. If too much sewage is discharged into a river, the amount of dissolved oxygen cannot keep pace with it. The helpful bacteria die, and the sewage is not broken down. In a number of places in the UK, this is happening. In busy estuaries like the Mersey, untreated sewage is a serious problem. Many of our rivers and estuaries are polluted. Water that contains substances which are harmful to health is *polluted*.

Safe Water

Our drinking water is safe. We take it for granted. It was not always so. In the nineteenth century, the diseases of cholera, typhus, dysentry, turberculosis and smallpox were common. These diseases are carried through drinking water. Now that we have sewage works to treat the sewage and water works to filter and chlorinate our water supply, these diseases have been wiped out.

There are still many parts of the world where people do not enjoy safe water. People in many African and Asian countries have to walk miles every day to the nearest river or well to collect water. Often, rivers are used as toilets as well as for drinking. Many diseases are waterborne. Typhus, cholera, river blindness, trachoma, bilharzia and diarrhoea are all carried by polluted water.

Figure 8.8   This is their water supply

Experiment 8.1

Which kind of soil retains the most water?

After rain, some soils dry out quickly, and others retain their water for some time. In this experiment, you are asked to plan and carry out an investigation to find out which kind of soil is the best at retaining water.

1. You will need samples of clay, sandy soil, loamy soil and any others you can find.

2. You are supplied with filter funnels, filter paper, glass wool, beakers, measuring cylinders, a top-loading balance and any other normal laboratory apparatus you need.

3. Write a brief account of what you plan to do to find the answer to the question. Say how you will make sure that the test is a *fair* test. Show the plan to your teacher, and, if he or she approves, put it into action.

8.5 Soap

One reason we need water is for washing. We keep a number of soap and detergent manufacturers in business. Soap is made by a process called *saponification*. Fats and strong alkali are boiled together (as in Experiment 8.2). The products are glycerol and a soap.

$$\text{Fat} + \text{Alkali} \longrightarrow \text{Glycerol} + \text{Soap}$$

One soap is sodium hexadecanoate. It is the sodium salt of hexadecanoic acid. Figure 8.9 shows a model of it. Count up; are there 16 carbon atoms? This is where the name comes from (*hexa* = *six*; *deca* = *ten*).

Figure 8.9   A model of a soap

The problem with washing is that grease and water do not mix. When you wash your hands with soap, the hydrocarbon part, ——$C_{17}H_{35}$, is attracted to the grease on your hands. The ——CO_2Na part is attracted to the water. The soap makes a bridge between the grease on your hands and the water. The grease floats away in the water, taking dirt with it (see Figure 8.10).

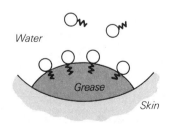

The water-hating tails of soap particles attach themselves to grease. The water-loving heads stick into the water.

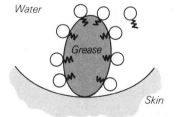

A blob of grease is squeezed away from the skin so that the water-loving heads can be surrounded by water. It floats away into the water.

Figure 8.10   How soap works

Experiment 8.2

To make a bar of soap

1. Put 25 g of a mixture of 25% vegetable oil and 75% dripping with 0.25 g of bar soap into an evaporating basin.

2. Set the evaporating basin over a beaker of water as shown in Figure 8.11 (a), and boil the water.

3. Gradually add 10 cm³ of 10% sodium hydroxide solution, stirring all the time.

4. Continue to heat and stir for about thirty minutes, until the whole mass in the dish has become emulsified and the mixture stiffens and sticks to the rod.

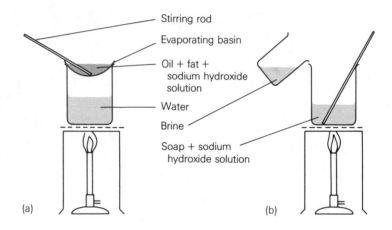

Figure 8.11   Making soap   (a) Saponification   (b) Salting out

5. Cool. Scrape into a 250 cm³ beaker, and add 30 cm³ of hot water.

6. Heat over a low flame, with stirring, for 30 minutes. When the contents of the beaker appear to be a thick, even paste, add saturated brine, as in Figure 8.11 (b). Keep stirring until the soap will break quickly and evenly from the surface of a spatula dipped into it.

7. Allow to stand overnight. The soap separates as a solid layer on top of the brine. The action of the brine in making the soap solidify is called salting out. Remove the soap and dry it with a paper towel. Add perfume and colouring if you wish, and shape the soap into a bar. Leave it to dry.

8. Do three tests with the soap you have made.
    (a) Wash your hands with it – not your face, as the soap may still contain some alkali.
    (b) Shake a piece of soap in a test tube with distilled water.
    (c) Shake a piece of soap in a test tube with tap water.

Soap lathers easily in distilled water, but sometimes it is hard to get a lather in tap water. We call this sort of tap water **hard**. Distilled water is described as **soft** water. It contains very little dissolved matter. Tap water contains dissolved salts, which make it hard. They waste soap by turning it into a **scum**.

Soaps are sodium salts of acids. They react with calcium and magnesium salts in water to form insoluble calcium and magnesium soaps. These insoluble salts are the scum that forms.

| Sodium soap (soluble) | + | Calcium sulphate (in hard water) | $\longrightarrow$ | Sodium sulphate (soluble) | + | Calcium soap (insoluble scum) |

When all the calcium and magnesium salts have been used up in forming scum, the soap can do its job. Hard water wastes a lot of soap. It also makes a bad job of laundry because scum sticks to the clothes. Removing the calcium and magnesium salts, that is, *softening* the water makes it possible for soap to do a better job.

8.6 Methods of Softening Hard Water

Temporary Hardness

Boiling. Temporary hardness is hardness which can be removed by boiling. This type of hardness is caused by calcium and magnesium hydrogencarbonates (see p. 124). When temporarily hard water is boiled, the hydrogencarbonates in it decompose.

Calcium hydrogencarbonate $\longrightarrow$ Calcium carbonate + Carbon dioxide + Water

$$Ca(HCO_3)_2(aq) \longrightarrow CaCO_3(s) + CO_2(g) + H_2O(l)$$

When calcium hydrogencarbonate decomposes, a deposit of insoluble calcium carbonate forms, and carbon dioxide is given off. The water is now soft. You can see a deposit of calcium carbonate on the inside of your kettle. We call it **scale** or **fur**. When a thick layer of scale builds up inside a hot water pipe, it can cause a blockage. When scale builds up inside a radiator, it can make the radiator heat less efficiently.

Permanent Hardness

Permanent hardness is hardness which cannot be removed by boiling. It is caused by the presence of calcium and magnesium chlorides and sulphates, which are not decomposed by heat. Various methods are used to soften permanently hard water. These methods also soften temporarily hard water.

Distillation. Distillation will remove all dissolved matter. This is too expensive a method for domestic and industrial use.

Washing soda. Washing soda and bath salts are sodium carbonate crystals, $Na_2CO_3.10H_2O$. They soften water by removing calcium and magnesium as insoluble carbonates.

| Calcium sulphate (in hard water) | + | Sodium carbonate (washing soda) | $\longrightarrow$ | Calcium carbonate (insoluble) | + | Sodium sulphate (in solution) |

$$CaSO_4(aq) + Na_2CO_3(aq) \longrightarrow CaCO_3(s) + Na_2SO_4(aq)$$

Exchange resins. There are materials called exchange resins which will soften water. Permutit® is an exchange resin which will exchange calcium and magnesium compounds for sodium compounds. When hard water trickles through a column of Permutit®, as shown in Figure 8.13, an exchange takes place.

| Calcium sulphate (in water) | + | Sodium permutit (solid resin) | $\longrightarrow$ | Sodium sulphate (in water) | + | Calcium permutit (solid resin) |

The water which leaves the Permutit® has lost its calcium and magnesium compounds, and is soft water. You can use Permutit® in Experiment 8.3.

Figure 8.12   Scale in a hot water cylinder

Experiment 8.3

To compare methods of softening tap water

1. **Tap water.** Measure 20 cm³ of tap water in a measuring cylinder, and pour it into a boiling tube. Add soap solution drop by drop from a teat pipette, with shaking, until a lather is formed which will last for ten seconds. Record the number of drops.

    (Another method is to add soap flakes and count the number of flakes.)

2. **Boiling.** Boil some tap water in a beaker for five minutes. Cool. Take 20 cm³ in a measuring cylinder, tip into a boiling tube and again find the number of drops of soap solution needed to give a lather. Write down the number.

A NEW FIRST CHEMISTRY COURSE

3. **Washing soda.** Measure 20 cm³ of tap water from a measuring cylinder into a boiling tube. Add a crystal of washing soda, sodium carbonate, and shake to dissolve it. Again, find the number of drops of soap solution needed to give a lather.

4. **Distillation.** Distil some tap water as shown in Figure 2.7. Test 20 cm³ of the distillate with soap solution as before.

5. **Permutit®.** Pass some tap water through a model of an industrial water softener (see Figure 8.13). Run tap water in slowly so that a slow trickle of water comes out of the bottom of the water softener. Collect it in a beaker, and measure 20 cm³ of it from a measuring cylinder into a boiling tube. Test with soap solution as before.

6. Now compare your results. Which samples of water needed the least soap solution to form a lather? Is there a high content of calcium and magnesium hydrogencarbonates (which are removed by boiling) in tap water? Is the water still hard when these compounds have been decomposed? Which method of softening worked best for you? You can find the ratio of temporary to permanent hardness from your results:

If number of drops of soap solution needed by tap water $= a$,
and number of drops of soap solution needed after boiling $= b$,

$$\text{permanent hardness} = b$$
$$\text{temporary hardness} = a - b$$

Then
$$\frac{\text{temporary hardness}}{\text{permanent hardness}} = \frac{a - b}{b}$$

There will be different ratios in different parts of the country.

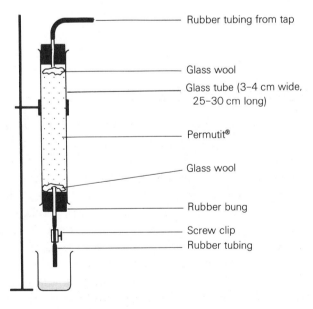

Figure 8.13   A model water softener

Detergents

Soapless detergents are made from petroleum oil and concentrated sulphuric acid. They clean in a manner similar to soaps: they form a bridge between grease and water. The difference is that detergents do not form scum even in hard water because the calcium and magnesium salts of soapless detergents are soluble.

People buy more soapless detergents than soaps. Before detergents became available, many houses had water softeners. These are commercial versions of the apparatus shown in Figure 8.13. Now, people prefer to use detergents which will work in hard water. In 1948, detergents had 10 per cent of the cleaning market; in 1953, they had 50 per cent; now, detergents have 80 per cent of the market.

There is a disadvantage to detergents. They pollute water. Phosphates are added to detergents. They make water alkaline, which makes the removal of grease easier. They also combine with calcium and magnesium compounds. When detergents are discharged into rivers and lakes, phosphates pollute the water. You can read about this in Section 8.8.

Advantages of Hard Water

For drinking, hard water is better than soft water. The calcium compounds in it help to form strong teeth and bones.

For some industrial uses, hard water is preferred. The brewing industry and the leather industry both use hard water.

8.7 Stalactites and Stalagmites

You have read how, in limestone regions, water contains dissolved calcium hydrogencarbonate. In an underground cavern, there will be a slow trickle of water running from the roof to the floor. Imagine a drop of water becoming isolated from the main stream. With air all round it, the water evaporates, and the salt dissolved in the drop of water is left behind as a tiny speck of calcium carbonate. Another drop of water evaporates, and another speck of calcium carbonate is formed. Over a period of hundreds of years, the specks of calcium carbonate accumulate and form a cone hanging down from the roof. This is called a **stalactite**. If evaporation takes place from the floor of the cave, a pillar can build up slowly from the floor. This is called a **stalagmite**. In some places, stalactites and stalagmites have met to form complete pillars (see Figure 8.14).

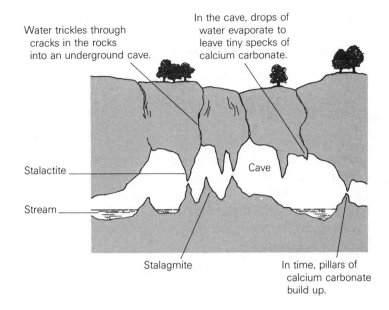

Water trickles through cracks in the rocks into an underground cave.

In the cave, drops of water evaporate to leave tiny specks of calcium carbonate.

Stalactite

Cave

Stream

Stalagmite

In time, pillars of calcium carbonate build up.

Figure 8.14   The formation of stalactites and stalagmites

Experiment 8.4

Which soap is the best?

For this experiment, you need tablets of different brands of soap. In each part of the experiment, you are asked to think up your own tests. When you have a plan, discuss it with your teacher before you go any further.

1. **Which kind of soap lathers most easily?**

    How can you make a fair test? You will have to give each tablet the *same* treatment, the same kind of agitation with water at the same temperature and keeping the rate of flow of water over the soap the same.

2. **Which soap lasts longest?**

    How can you make a fair test? You may be able to think up a test which involves weighing. You will have to be sure that each tablet gets the same treatment.

3. **Which soap is the least 'slushy'?**

    Some soaps become very slushy when you put them into a wet soap dish. Can you devise a test for 'slushing'? Which soap gives the most gel in the soap dish?

4. There are other things about soaps which influence people's choice. What are they? Are there any more tests which you should do?

Experiment 8.5

Which is better, soap or detergent?

1. Stain a piece of cotton cloth with coffee, and let it dry.

2. Obtain a soap powder (e.g. Lux®) and a soapless detergent (e.g. Persil®).

3. Can you think up a fair way of testing to find out which of the two, soap or detergent, is better at removing the coffee stain? You will need to decide the amounts of soap and detergent to use. You must decide the kind of washing treatment you will give the cloth – how long, with how much agitation and at what temperature. The treatment must be the same for both soap and detergent. Ask your teacher's advice before you do the test you have planned.

4. Repeat your test with other common stains, e.g. gravy, tea, cocoa, curry, make-up, shoe polish, felt-tip pen, coke and others.

5. Extend your work to other fabrics, e.g. wool and nylon.

Experiment 8.6

Which is the best shampoo?

A shampoo is a solution of a detergent in water with antiseptics, colouring and perfume. You are asked to design experiments which will help you to compare shampoos. Before you do the experiments you have in mind, be sure to ask your teacher to check your plans.

1. **How much water does a shampoo contain?**

   To find out, you will have to weigh some shampoo and then heat it gently to drive off water. About 5 cm³ of shampoo should be enough. What will you weigh it on? You must not heat strongly or the shampoo will decompose. How will you provide gentle heating?

   What weighings will you need to do?

   How will you calculate the percentage of water in the shampoo?

2. **Which shampoo is best at removing grease?**

   You could wash a grease-coated lock of hair or a grease-coated hank of wool. Think up a fair method of comparing the action of different shampoos. Draw up a plan before you start testing.

3. **Does a 'protein shampoo' really mend split ends?**

   You can find out with the help of a microscope. You will need to look at a lock of hair under a microscope before and after washing. Write out a plan before you start work. Do not forget about doing a control experiment.

4. Are there any other factors which people consider before choosing a shampoo? If there are, how can you test these other factors?

5. Can you rank the different shampoos you have tested? Put them in order for test 1, test 2, test 3 and also price.

Experiment 8.7

Which is the best detergent?

Which detergent will emulsify the most oil?

1. Make a solution of 5 cm³ of detergent in 2.5 dm³ of water.

2. Take 10 cm³ of the solution of detergent. Take 1 drop of cooking oil on the end of a glass rod, and add it to the detergent. Stopper the test tube, and shake it for 10 seconds.

3. Add another drop of oil, stopper and shake again.

4. Repeat until you can see a very thin layer of oil under the froth. Write down the total number of drops of oil.

5. Enter the price of the detergent in the table.

Detergent	Number of drops of oil emulsified	Price of detergent (p per 100 g)	Number of drops of oil / Price in p per 100 g
Twinkle Kleenov			

6. When you have tested as many detergents as you can, rank them: best emulsifier ... good ... medium ... poor ... worst ...

   Can you choose a best buy? The best buy will have the highest ratio of number of drops of oil emulsified/price per 100 g of detergent.

8.8 Polluted Water

Water which contains substances which are harmful to health is described as *polluted*. There are a number of ways in which pollutants can enter water.

Sewage

Large quantities of sewage are discharged untreated into estuaries and coastal waters. In some holiday resorts, bathing beaches have been contaminated by sewage washing ashore. In many estuaries, catches of fish have fallen because decaying sewage uses up dissolved oxygen. Section 8.4 explained the importance of dissolved oxygen.

Thermal Pollution

Many industries use water as a coolant. Electricity power stations use large volumes of water. They take it from a river, circulate it round the plant and return it to the river. In many cases, this does no harm. Oxygen is, however, less soluble in hot water than in cold water. Warming a river decreases its content of dissolved oxygen. If the river is already short of dissolved oxygen, this may be serious.

Acid Rain

Acid rain pollutes lakes and forests. You read about this in Chapter 5.

Agricultural Pollution

Farmers use fertilisers to increase their yields of cereals, fruits and vegetables. Fertilisers contain nitrates (nitrogen compounds) and phosphates (phosphorus compounds). Often, a crop receives more fertiliser than it can use. Then, fertiliser washes deep into the ground before the plants have had time to absorb it. Fertiliser finds its way into lakes and rivers and into ground water.

If lake water becomes rich in nitrates and phosphates, it becomes more fertile for plants to grow in. The lake becomes covered by a layer of algae and clogged with weeds. When these plants die, the process of decay uses up dissolved oxygen. Lakes are usually a good source of fish. In a lake which is short of dissolved oxygen, the fish die. This accidental enrichment of lakes and rivers is called **eutrophication**. A eutrophic lake or river is no use for drinking, swimming, boating or fishing.

Fertiliser finds its way into ground water. This is the water held underground in layers of porous rock. In the UK, a third of our drinking water comes from ground water. Many people are worried about the nitrates in ground water. The reason is that, at a high enough level, nitrates in drinking water can result in damage to haemoglobin, the red pigment in blood. The nitrate content of drinking water is increasing. It is not yet at the level which the World Health Organisation says is dangerous.

Using too little fertiliser, farmers lose money because their crops are poor. Using too much fertiliser, they waste money because some of the fertiliser is not used by the crop. The solution to the problem is to apply the right amount of fertiliser at the right time, during the plants' growing season. Chemists in agricultural research laboratories are working out better methods of matching the fertiliser applied to the plants' ability to absorb it.

Detergents

Detergents contain phosphates which act as 'brighteners'. When detergents are discharged into lakes and rivers, they stimulate plant growth in the same way as fertilisers.

Industrial Pollution

Industries use the river flowing past the factory as a useful method of carrying away waste products. Dirt, oil, acids, alkalis and poisonous salts are poured into rivers. The industrialists hope that the rivers will deal with these products by diluting them to harmless levels. This happens in some cases. Many rivers have so many factories discharging into them that they cannot cope with the pollutants. In industrial areas, rivers are dirty brown streams into which no light can penetrate and in which no fish or plant can live. In some rivers, the levels of poisonous metal salts are building up.

Oil Tankers

Oil tankers carry enormous loads of crude oil from oil fields to refineries. A large modern oil tanker is as long as four football pitches and carries about 250 000 tonnes of crude oil.

Figure 8.15   An oil tanker

There are three ways in which oil may be spilt:

- accidental spillage during loading and unloading
- collisions with other vessels and groundings
- illegal washing out of tanks at sea.

Tankers are forbidden by law to wash out dirty tanks at sea. Nevertheless, some tankers do this on their way from a port where they have discharged a load to a port where they will pick up a fresh cargo. By doing this job at sea, instead of in port, they save time.

The problem of oil spillage is a serious one. Oil washes ashore and pollutes the coast. This happened in 1967 when the *Torrey Canyon* sank. Oil polluted beaches in Cornwall and the Scilly Isles. It happened to the coast of Brittany in 1978 when the *Amoco Cadiz* had an accident off the coast of France. These areas are popular resorts. They lost a lot of income because holiday makers do not want to spend their time on beaches covered with oil.

Most of the oil floats on the surface of the sea. Slowly, air oxidises it to carbon dioxide and water. Bacteria help to decompose it. While the oxidation is going on, dissolved oxygen is being used up. Fish, unable to obtain enough oxygen, die. Fishermen lose money as their catches suffer.

Seabirds dive for food. They can dive from a clear patch of water and, on coming up for air, find themselves in a patch of oil. Oil sticks their feathers together so that they cannot fly. It ruins their insulation from the cold weather. Unable to fly or to keep warm, sea birds die in thousands.

Figure 8.16 A sea bird damaged by oil

What Do You Know about Pollution?

1. Why are there more lakes than rivers suffering from *eutrophication*?

2. What effect does acid rain have on (a) trees (b) lakes? (See Chapter 5 if you need help.)

3. (a) Why does a river need plenty of dissolved oxygen?

   (b) How does sewage use up this oxygen?

4. (a) Why is a eutrophic lake short of oxygen?

   (b) What does a eutrophic lake look like?

   (c) Why is it unpleasant to swim in?

   (d) Why can the water not be used for drinking?

5. If detergents are discharged into a lake, they cause pollution. What sort of pollution results? What can you see in a lake which is affected in this way?

6. (a) What is *ground water*?

   (b) How can it become polluted?

   (c) Why would polluted ground water be a serious matter?

7. (a) What is the advantage of using a large amount of fertiliser on a growing crop?

   (b) What is the disadvantage?

   (c) How can a balance be found between the advantage and the disadvantage?

8. What do people mean by saying that heat can cause water to become polluted?

9. (a) How can detergents pollute water?

   (b) Why do people prefer detergents to soaps?

10. (a) What is the harm in oil tankers washing out their tanks at sea?

(b) What can be done to prevent it? (It is already illegal.)

8.9 Splitting up Water

One method of splitting up a compound is to pass a direct electric current through it. Experiment 8.8 shows what happens when you do this with water.

Experiment 8.8

To pass a current from a battery or a labpack through water

Figure 8.17 shows an apparatus which can be used

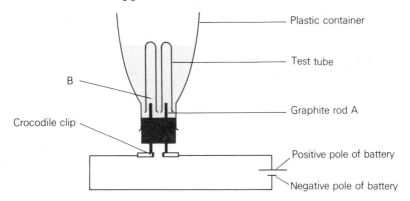

Figure 8.17   Passing a direct electric current through water

1. Fit graphite rods, which are called electrodes because they carry an electric current, through a rubber bung into a container. The one shown is a plastic juice bottle with 10 cm cut off the bottom. Attach crocodile clips to leads coming from a 6 V battery or a labpack adjusted to give 6–12 V.

2. Fill the container with distilled water. Add a few drops of dilute sulphuric acid. This helps the water to conduct electricity. Connect the crocodile clips to the graphite electrodes.

3. If you see bubbles of gas forming at an electrode, fill a test tube with water, put your thumb over the open end, and invert it over the electrode. It is better to clamp the test tube, not to rest it on the rubber bung.

4. Test any gas collected. Put your thumb under the test tube, remove it from the water, put a lighted splint into the test tube.

5. Was a gas formed at A? How did it react to a lighted splint? Was a gas formed at B? How did it react to a lighted splint?

Copy and complete the equation:

Water $\longrightarrow$ _____ + _____

___ $H_2O(l)$ $\longrightarrow$ _____ + _____

If you have made the tests described in Experiment 8.8, you have found that the gas at A, which is connected to the positive end of the battery, is oxygen. The gas at B, which is connected to the negative end of the battery, burns with an explosive 'pop'. It is another element, and was given the name *hydrogen*, which means *water-maker* in Greek. Thus water has been split up by electricity into the elements hydrogen and oxygen. The process of splitting up by electricity is called *electrolysis*. Water is a compound of hydrogen and oxygen. You may have noticed that the volume of hydrogen produced is twice that of oxygen. Experiments on these lines have shown that the formula of water is H_2O.

8.10 Tests for Water

There are many colourless liquids. You cannot assume that any colourless, odourless liquid you see is water. You need to make tests. There are many chemical reactions in which water plays a part. Here are two colourful ones, which will tell you whether a liquid contains water. They do not tell you whether the liquid is pure water or not.

Test	Result
Add the liquid to *anhydrous* copper sulphate (copper sulphate which contains no water). This has the formula $CuSO_4$ and is *white*.	A compound of copper sulphate and water called a *hydrate* is formed. It has the formula $CuSo_4.5H_2O$, and it is *blue*.
Add the liquid to anhydrous cobalt chloride, $CoCl_2$, which is *blue*.	If the liquid contains water, the hydrate $CoCl_2.5H_2O$ is formed. This is *pink*.

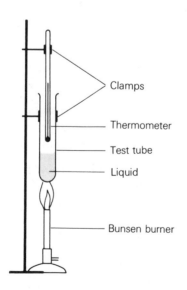

Figure 8.18   Finding the boiling point of a liquid

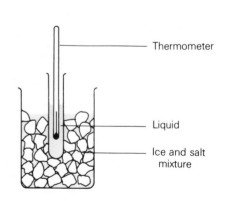

Figure 8.19   Finding the freezing point of a liquid

Having found out that the liquid contains water, you can now test to see whether it is pure water. Water has boiling point 100 °C, freezing point 0 °C, and density 1 g/cm^3. To find the boiling point, set up the apparatus shown in Figure 8.18. When the liquid boils, record the temperature shown by the thermometer.

To find the freezing point, you set up the apparatus shown in Figure 8.19. The mixture of ice and salt cools the water in the test tube. The temperature recorded on the thermometer falls and then remains constant when the water in the test tube starts to freeze, and stays constant all the time the liquid around it is freezing.

Questions on Chapter 8

1. (a) Name three processes which send water vapour into the atmosphere.

    (b) What is the purest form of natural water?

    (c) What is *ground water*? Why can ground water be used as drinking water after only a small amount of treatment?

    (d) Why does river water need much more purification before it can be drunk?

    (e) There are two main processes in the treatment which river water receives at the water works. Describe them.

2. (a) How is the waste water treated at the sewage works?

    (b) Why is some waste water discharged untreated into rivers?

    (c) What is the danger in doing this?

3. (a) Explain why washing a pair of socks in soap and water is better than washing them in water alone.

    (b) In what way are soapless detergents better than soaps? What disadvantage do soapless detergents have?

4. (a) What is meant by *hard* water?

    (b) Why does using bath salts make your bath water soft? Point out two advantages of using bath salts.

5. (a) Why does scale form inside a kettle?

    (b) Why do stalactites form in underground caves?

6. Say what damage can be caused by oil spillage at sea

    (a) to the sea

    (b) to the shore

    (c) to marine birds.

7. When an industry uses water for cooling, the temperature of the water is raised. This is described as *thermal pollution*. What harm can it do?

8. How can water become polluted by (a) fertilisers (b) detergents? What does such polluted water look like?

9. Write down the missing words in this passage.

Temporarily hard water contains _____, which when heated becomes _____. This reaction occurs naturally in limestone caves, where _____ is deposited as _____ on the roof and _____ on the floor. Temporary hardness can be removed by _____. Permanent hardness is caused by salts such as _____. Hard water can be softened by adding _____.

10. (a) What do you see when soap is shaken with hard water? Explain the chemical reaction that occurs. Why does this reaction not happen in soft water?

(b) What happens when detergents are used in (i) soft water and (ii) hard water? What advantage do detergents have over soaps? What pollution is caused by detergents?

11. (a) For what purposes is hard water better than soft water?

(b) Why does scale form in hot water pipes but not in cold water pipes?

(c) What would happen if a piece of stalactite were dropped into dilute hydrochloric acid?

12. Which lathers more easily with soap flakes, Glasgow water or Cardiff water? Think up a fair method of testing. List carefully all the things that you will have to keep the same in tests on the two kinds of water.

Crossword on Chapter 8

Trace or photocopy this grid (teacher, please see note at the front of the book), and then fill in the answers.

Across

1 Turn into a liquid (8)
5 Plenty of water flows in this (5)
6 Smallest (5)
9 Symbol for tantalum (2)
10 Women's Army (1,1)
11 Soaps and detergents compete as _____ in the market (6)
15 Symbol for arsenic (2)
16 Do this to water to get hydrogen and oxygen (11)
17 Same clue as 9 across
18 You won't get scum with this (9)
19 These break down sewage (8)
22 The process of transferring water from the atmosphere to the ground (7)

Down

1 This disinfects water (8)
2 See 10 down
3 The method of obtaining the purest water (12)
4 Soap removes this from skin (6)
7 This grows in an underground cavern (10)
8 A story (4)
10, 2 down. This softens water (7,4)
12 A doctor for animals (3)
13 Clean thoroughly (5)
14 This fruit contains tartaric acid and citric acid (5)
18 Polluted lakes which have lost their fish are described as _____ lakes (4)
20 Symbol for iridium (2)
21 Automobile Association (1,1)

9 Metals

9.1 We Use Metals

Of the 92 elements that are found in the Earth's crust, 70 are metals. We find important uses for about half of these. In addition, we use **alloys**. These are not elements; they are mixtures of metallic elements (and sometimes non-metallic elements too). Alloys are often stronger than pure metals. Why is aluminium alloy used for the manufacture of artificial limbs? Why is gold used for the contacts in the electric circuits in space craft? Why is mercury chosen as the basis of the dental amalgam used for filling teeth? Why do Rolls-Royce choose stainless steel for the radiator grilles on their cars? Why is steel the alloy used to make tankers? Why are steel girders galvanised with zinc? You looked at the physical characteristics of metals in Chapter 4. In this chapter, you will look at the chemical reactions of metals. When you have done this, you will be able to answer these questions and to see why different metals and alloys are chosen for different purposes. You can read more about metals and alloys in *Extending Science 5: Metals and Alloys* by E. N. Ramsden (ST(P)).

Figure 9.1   The 'silver lady' and radiator grille are stainless steel

9.2 Metals which React with Water

In this section, we are going to study the reactions which take place between water and metals. One you have already met is the rusting of iron (Chapter 5). There are other metals which react with water.

Demonstration Experiment 9.1

To study the reaction of water with sodium and potassium

> **A safety screen should be used for this demonstration.**
> **Observers should wear safety glasses.**

1. Sodium is so reactive that it is kept under oil to prevent air and water from reaching it. Cut a piece of sodium the size of a pea.

2. Drop the piece of sodium into a trough of water. Observe all that happens, and test the water with litmus.

3. To test the idea that heat is generated, see what happens when sodium is not free to move. Float a piece of filter paper in the trough, and drop a piece of sodium on to it. The sodium cannot move, but water can reach it through the filter paper.

   The sodium bursts into flame, and burns with a yellow flame – or is it a gas formed during reaction that burns?

4. To test for gas, position a Pyrex tube around the sodium as shown in Figure 9.2. When the sodium has finished reacting, bring a lighted taper to the top of the tube.

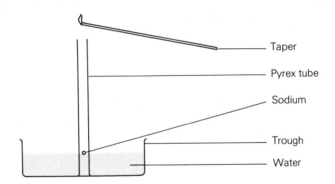

Figure 9.2   Testing the gas formed when sodium reacts with water

5. Repeat steps (1) and (2) with potassium.

6. Answer these questions about what you have seen.
   (a) Is sodium a hard metal or a soft metal?
   (b) What does a freshly cut surface of sodium look like?
   (c) Why does it soon tarnish?
   (d) Why is sodium kept in a bottle of oil?
   (e) Does sodium float or sink in water? Is sodium denser or less dense than water?
   (f) Think carefully. What evidence have you that heat is given out when sodium reacts with water?
   (g) What is it that causes sodium to move around?
   (h) Which gas is formed in the reaction?
   (i) What happens to the water in the trough?

(j) What evidence have you seen that potassium is more reactive than sodium?

(k) What is the colour of (i) the sodium flame and (ii) the potassium flame?

7. The equations for the chemical reactions that have taken place are:

Sodium + Water $\longrightarrow$ Hydrogen + Sodium hydroxide solution

$2Na(s) + 2H_2O(l) \longrightarrow H_2(g) + 2NaOH(aq)$

Potassium + Water $\longrightarrow$ _____ + _____

$2K(s) + 2H_2O(l) \longrightarrow$ _____ + _____

Experiment 9.2

To study the reaction of calcium with water

1. Take a 250 cm^3 beaker three quarters full of water. Add a piece of calcium. What happens?

2. Invert a test tube full of water over the piece of calcium. Collect the gas formed, as in Figure 9.3 (a). Put your thumb over the open end of the test tube while the tube is under water, as in (b). Open the tube at a flame, as in (c). What happens?

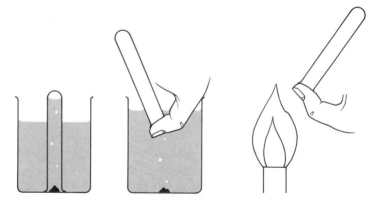

Figure 9.3   The reaction between calcium and water

3. When the calcium has reacted, what is left in the beaker? Can you name this substance?

4. Filter the contents of the beaker.

5. Take a portion of the filtrate in a test tube. Add two drops of litmus solution. What happens?

6. Take a second portion of the filtrate and blow into it through a straw. What happens? Can you name this solution?

7. Explain why

(a) calcium is kept in a closed bottle, and

(b) it is not kept under oil as sodium is.

8. (a) Name the filtrate formed in step (4). Explain how you identified it.

   (b) Name the gas formed in step (2)

   (c) Complete the word equation,

   Calcium + Water ⟶ _____ (g) + _____ (aq)

   and write a balanced chemical equation,

   Ca(s) + _____ ⟶ _____ + _____

Experiment 9.3

To find out whether magnesium reacts with water

1. Sprinkle a spatula measure of magnesium powder into a 400 cm³ beaker full of water. Cover the magnesium powder with a filter funnel, as shown in Figure 9.4. Invert a test tube of water over the neck of the filter funnel.

2. Leave to stand. Test any gas formed by opening the tube at a flame.

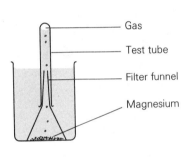

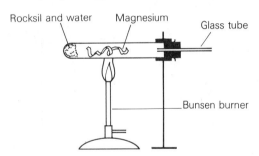

Figure 9.4   Reaction of magnesium and water

Figure 9.5   The action of magnesium on steam

Experiment 9.4

To find out whether magnesium reacts with steam

1. Place loosely packed rocksil in a hard glass test tube to a depth of 3 cm. Add, by a teat pipette, as much water as the rocksil will hold (about 2 cm³).

2. Coil a piece of magnesium ribbon, and put the coil into the middle of the tube. Fit a glass tube and clamp as shown in Figure 9.5.

3. Wearing safety glasses, heat the magnesium. Enough heat will reach the water and turn it into steam.

4. When you see that a reaction is occurring between the magnesium and the steam, apply the Bunsen flame to the end of the glass tube. Make a note of what you observe.

5. When the reaction is over, tip out the product formed by magnesium and steam. Add water, warm, and test with universal indicator.

6. (a) How long does it take before Experiment 9.3 gives enough hydrogen to 'pop'?

   (b) How long does it take to obtain hydrogen in Experiment 9.4?

   (c) What is the reason for the difference?

   (d) Complete these equations for the reaction between magnesium and steam:

   $$\text{Magnesium} + \text{Steam} \longrightarrow \underline{\hspace{2cm}} + \text{Magnesium oxide}$$

   $$\text{Mg(s)} + \text{H}_2\text{O(g)} \longrightarrow \underline{\hspace{1.5cm}} + \underline{\hspace{1.5cm}}$$

Experiment 9.5

To study the action of steam on iron and on zinc

1. Place loosely packed rocksil in a hard glass test tube to a depth of 3 cm. Add with a teat pipette as much water as the rocksil will hold.

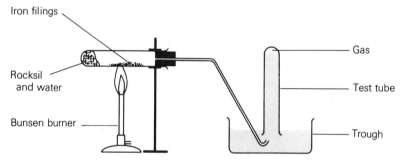

Figure 9.6   The reaction of steam and iron

2. Spread a spatula measure of iron filings in the test tube. Attach a delivery tube, and position a test tube to collect any gas formed, as shown in Figure 9.6.

3. Heat the part of the test tube containing iron. Do not heat the rocksil directly: by moving the flame to and fro, boil the water in the rocksil and keep the metal hot. In this way, a gentle flow of steam is passed over the hot metal.

4. Collect any gas formed. With your thumb over the end of the test tube, remove it and open it at a flame. What happens?

5. As soon as you stop heating, lift the delivery tube out of the trough to avoid 'sucking back'. This is the tendency of water to travel up the delivery tube when the pressure of hot gases in the Pyrex tube drops. Cold water reaching the hot tube will crack it.

6. Describe the change in appearance of the iron. Copy and complete the equation

   $$\text{Iron} + \text{Steam} \longrightarrow \underline{\hspace{2cm}} + \underline{\hspace{1.5cm}}$$

7. Carry out the experiment in exactly the same way with zinc instead of iron.

Summary of the Reactions Between Metals and Water

Sodium

The reaction between sodium and water is vigorous. The hydrogen which is formed often bursts into flame. The flame is coloured yellow by sodium.

Sodium + Water $\longrightarrow$ Hydrogen + Sodium hydroxide

$$2Na(s) + 2H_2O(l) \longrightarrow H_2(g) + 2NaOH(aq)$$

Potassium

The reaction is violent. The hydrogen burns with a flame coloured lilac by potassium.

Sodium and potassium must be kept under oil to prevent water vapour in the air from reacting with them. They are called **alkali metals** because their hydroxides, NaOH and KOH, are strong alkalis.

Calcium

Calcium reacts readily with cold water. Hydrogen and the alkali calcium hydroxide (limewater) are formed.

Calcium + Water $\longrightarrow$ Hydrogen + Calcium hydroxide

$$Ca(s) + 2H_2O(l) \longrightarrow H_2(g) + Ca(OH)_2(aq)$$

Magnesium

Magnesium reacts very slowly with cold water but burns in steam.

Magnesium + Water $\longrightarrow$ Hydrogen + Magnesium hydroxide
(cold)

$$Mg(s) + 2H_2O(l) \longrightarrow H_2(g) + Mg(OH)_2(aq)$$

Magnesium + Steam $\longrightarrow$ Hydrogen + Magnesium oxide

$$Mg(s) + H_2O(g) \longrightarrow H_2(g) + MgO(s)$$

Iron and Zinc

Iron and zinc react with steam to form hydrogen and the metal oxide. Iron forms the blue-black oxide Fe_3O_4. This is different from rust (Fe_2O_3). Zinc forms the oxide ZnO, which is yellow when hot and white when cold.

9.3 Metals which React with Dilute Acids

Most metals react with acids. Acids contain hydrogen, and they part with it more readily than water does.

Experiment 9.6

To investigate the action of dilute hydrochloric acid on metals

1. Place a row of test tubes in a rack. Half fill them with dilute hydrochloric acid.

2. Add a piece of metal to each, labelling the tubes as you do so. *Do not* use the metals which react with cold water (sodium, potassium and calcium) as they will react too vigorously with acid. Use the metals magnesium, zinc, aluminium, tin, iron, lead and copper.

3. Make a note of any changes which occur in the test tube. If a gas is given off, test it with a lighted splint.

4. If there is no reaction, try warming the test tube with a small flame.

5. Draw up a table of your results.

Reactions of metals with dilute hydrochloric acid

Metal	Is hydrogen given off?	Is the reaction fast or slow?	Appearance of the solution

6. Copy and complete the word equations:

    Zinc + Hydrochloric acid ⟶ _____ + _____

    Magnesium + Hydrochloric acid ⟶ _____ + _____

    Write the chemical equations for these two reactions.

7. Rank the metals which you have tested in order of reactivity, with the most reactive first.

Table 9.1 *Reactions of some common metals*

Element	Reaction with air	Reaction with cold water	Reaction with steam	Reaction with dilute acid
Potassium	Burn in air or oxygen	Vigorous	Dangerously fast	Dangerously fast
Sodium				
Calcium				
Magnesium		Slow	Fast	Fast
Zinc		No reaction	React	Fairly fast
Iron				
Tin	Form oxide when heated in air		No reaction	Slow
Aluminium				
Lead				No reaction (except oxidising acids)
Copper				

Aluminium does not show its true reactivity. The metal is surrounded by a layer of aluminium oxide which protects it from acid attack. This layer can be removed by the action of concentrated hydrochloric acid. Then the fresh aluminium underneath will show its true reactivity and place the metal just below magnesium in reactivity. Oxide films give incorrect results for other metals too. The correct order of reactivity is given in Table 9.2. Included are gold, silver, platinum and mercury, which are too expensive for us to experiment with but familiar to you outside the laboratory. These metals are less reactive than copper.

The beautiful appearance of silver and gold and the fact that they are chemically unreactive makes these metals so precious. This order, which places the metals in order of reactivity, is called the **reactivity series**. Hydrogen is included. Metals above hydrogen in the reactivity series will displace it from dilute hydrochloric and sulphuric acids. Metals below hydrogen will not displace it from dilute acids. Lead is just above hydrogen in the reactivity series, but in practice its reactions with dilute sulphuric and hydrochloric acids are too slow to be of use to us.

Table 9.2   *The reactivity series*

Potassium
Sodium
Calcium
Magnesium
Aluminium
Zinc
Iron
Tin
Lead
Hydrogen
Copper
Silver
Mercury
Platinum
Gold

When the information from chemical reactions is in doubt because the metal forms an oxide film, it is better to rely on information obtained by a physical method which measures the voltage needed to split up salts of the metals (as in Experiment 8.8 on the electrolysis of water).

Displacement Reactions

When we say that one metal is more reactive than another, we mean that it is more ready to form compounds. If a metal X is more reactive, that is more ready to form compounds, than a metal Y, then X should displace Y from a solution of a compound of Y:

$$X(s) + Y \text{ compound (aq)} \longrightarrow Y(s) + X \text{ compound (aq)}$$

To test this idea, take two metals widely separated in the reactivity series, such as magnesium and copper, and see whether magnesium will displace copper from a solution of copper sulphate. Experiment 9.7 gives details, and Experiment 9.8 extends this work.

Experiment 9.7

To find out whether magnesium will displace copper from a solution of copper sulphate

1. Take a test tube half filled with copper sulphate solution. Drop a piece of magnesium ribbon into it, and leave it to stand.

2. Observe any changes in the metal and in the solution.

3. Copy and complete the word equation:

    Magnesium + Copper sulphate $\longrightarrow$ _____ + _____

    Can you write a chemical equation for the reaction?

    Is magnesium above or below copper in the reactivity series?

Experiment 9.8

To put metals in order of reactivity by studying displacement reactions

1. Fill a test tube half full of a concentrated solution of a compound of one metal. Hang a strip of another metal in it, as in Figure 9.7. Leave it to stand for quarter of an hour.

2. Inspect carefully. Make a note of any changes in the metal and in the solution. Which metals have been displaced?

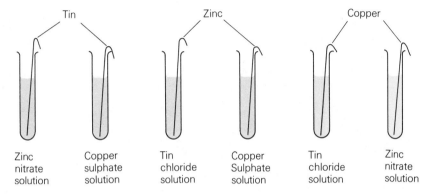

Figure 9.7   Displacement reactions of metals

3. Copy the table and fill in your results. What experiments would you have to do before you could complete the table?

Compound	Results obtained with the metal		
	Tin	Zinc	Copper
Zinc nitrate Copper sulphate Tin chloride			

4. Which of the three metals is the least reactive?

   Which of the three metals is the most reactive?

   Put the metals in order: most reactive > next > least reactive.

5. Iron is an important metal. Is it more reactive or less reactive than zinc? Sketch an experiment which you could do to find out. When you have your teacher's approval, do the experiment you have in mind. Then rank the four metals, tin, zinc, copper and iron, in order of reactivity.

6. Two more suggestions for you to try are:
   (a) copper + silver nitrate solution
   (b) zinc + lead nitrate solution

7. When you have done all the experiments, you will be able to rank in order of reactivity the metals:

   magnesium, copper, tin, iron, zinc, silver and lead.

9.4 Hydrogen

The reaction of a metal with a dilute acid gives a convenient way of preparing hydrogen. You can try it in Experiment 9.9.

Experiment 9.9

Laboratory preparation of hydrogen

1. Set up the apparatus shown in Figure 9.8.

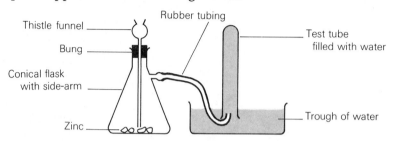

Figure 9.8   Apparatus for preparing hydrogen

2. Light some Bunsen burners on the side benches. Keep the Bunsens away from your apparatus.

3. Add 15 cm³ of dilute sulphuric acid. The acid must cover the bottom of the thistle funnel.

4. Add 1 cm³ copper sulphate solution. Stopper. Note what happens.

5. Collect the gas after it has been bubbling for half a minute.

6. Remove a test tube full of gas, closing the end with your thumb. Open it near a Bunsen flame. What happens?

7. Note the colour and smell of the gas. Collect two more test tubes of gas, and insert corks while the open ends are under water.

8. Is hydrogen denser or less dense than air? Put a test tube of dry air above a corked test tube of hydrogen, as shown in Figure 9.9 (a). Remove the cork, and hold the mouths of the two test tubes together while you wait ten seconds. Then take the tubes apart and immediately apply the ends to a flame. Where is the hydrogen now?

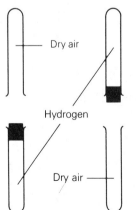

Figure 9.9   Is hydrogen denser or less dense than air?

9. Repeat (8), this time with the test tube of hydrogen on top of the test tube of dry air, as in Figure 9.9 (b).

What do you observe in the dry test tube after a mixture of hydrogen and air has burned in it?

10. Summarise your results by copying and completing this passage.

Hydrogen is _____ in colour and _____ in smell. The gas is soluble/insoluble in water. It is denser/less dense than air. When hydrogen burns, you hear _____ and on the sides of the container you see _____.

The equation for the reaction is

$$\text{Hydrogen} + \text{Oxygen} \longrightarrow \text{_____}$$

$$\text{_____} H_2(g) + \text{_____} O_2(g) \longrightarrow \text{_____}$$

The combustion is an example of two elements combining to form a compound. The elements are _____ and _____; the compound is _____.

Experiment 9.10

How fast do metals displace hydrogen?

Many metals displace hydrogen from acids. A reactive metal displaces hydrogen faster than a less reactive metal. One way of putting metals into their order of reactivity is to compare the speeds with which they displace hydrogen from acids.

Can you think up a method of finding the time taken for the reaction between a metal and an acid to produce 10 cm^3 (half a test tube) of hydrogen? You will have to think of a method of collecting the gas and measuring its volume. Then you can time the collection of 10 cm^3 of hydrogen from the reaction of zinc and hydrochloric acid. You will have to repeat the measurement with different metals and the same acid.

Write a plan of your experiment.

How will you collect hydrogen?

How will you measure 10 cm^3 of the gas?

How will you time the production of 10 cm^3 of the gas?

What factors will you have to keep the same in tests on different metals?

Show your plan to your teacher before you begin experimenting.

Tabulate your results:

Metal	Time taken for the production of 10 cm^3 of hydrogen
Zinc Iron Magnesium others	

Industrial Manufacture of Hydrogen

Petroleum oil. Crude oil gives more heavy fuel oil than we need and less petrol and aviation fuel than we need. The petroleum industry can make more petrol and aviation fuel by heating heavy fuel oil. In the process, hydrogen is formed.

Electrolysis. You saw how hydrogen is formed when water is *electrolysed* (split up by an electric current). In industry, brine (sodium chloride solution) is electrolysed to give three important products, hydrogen, chlorine and sodium hydroxide.

Industrial Uses of Hydrogen

Manufacture of ammonia. Hydrogen combines with nitrogen to form ammonia, NH_3. This is an important reaction because ammonia is used in the manufacture of fertilisers (see Figure 5.17). Ammonia is important for another reason. It can be oxidised to nitric acid, which is used in the manufacture of explosives, such as nitroglycerine and trinitrotoluene (TNT).

Reducing agent. Hydrogen is used as a reducing agent in the extraction of metals from their ores, for example,

$$\text{Hydrogen + Tungsten oxide} \longrightarrow \text{Tungsten + Water}$$

Tungsten is needed for the manufacture of electric light filaments.

Fuel. Hydrogen is used as a fuel. It burns to form the harmless product water. Space rockets carry liquid hydrogen for fuel and liquid oxygen for it to burn in.

Hydrogenation of oils. Vegetable oils, such as peanut oil, can be converted into hard fats, such as margarine, by reaction with hydrogen in the presence of nickel. The process is called **hydrogenation**.

Balloons. Figure 9.10 shows scientists from the Meteorological Research Station releasing a hydrogen balloon. The balloon will carry equipment up into the atmosphere. The balloon's instruments will record information about the atmosphere which the meteorologists (weather scientists) will study.

Figure 9.10   Scientists releasing a meteorological balloon

Questions on Chapter 9

1. In 1986, a fire broke out in the warehouse of a factory in Switzerland. As a result, 30 tonnes of pesticide were washed into the River Rhine. Alongside the warehouse was a storage shed for sodium. Its roof was punctured by drums of chemicals which had been thrown into the air by explosions in the burning warehouse. Firemen soaked all the buildings around the warehouse with water.
    (a) What would have happened if a drum of sodium had cracked and come into contact with water?
    (b) What harm do you think the pesticides did in the River Rhine?

2. The metal cobalt reacts with lead nitrate to form lead and cobalt nitrate.
    Cobalt does not react with zinc nitrate.
    (a) Place the three metals, cobalt, lead and zinc, in order of reactivity (most reactive first).
    (b) Say what you would expect to happen if (i) zinc were added to a solution of cobalt nitrate and (ii) lead were added to a solution of zinc nitrate.

3. You are given a sample of a solid element **X**.
    (a) Say what you would do to find out whether **X** is a metal.
    (b) If **X** is a metal, say what you would do to find its position in the reactivity series.

4. Explain the following.
    (a) Lead, not iron, is used for church roofs.
    (b) Iron, not lead, is used for machinery.
    (c) Copper, not iron, is used for water pipes.
    (d) Aluminium, not iron, is used for aeroplanes.
    (e) Aluminium, not iron, is used for soft drink cans.
    (f) Aluminium, not silver, is used for baking foil.
    (g) Stainless steel, not iron, is used for cutlery.
    (h) Stainless steel, not copper, is used for radiators.

5. Copy the flow chart, and fill in the missing information.

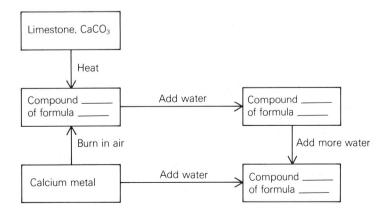

6. Explain the choice of the metal for the job.

   (a) Why is gold used in jewellery?

   (b) Why is mercury chosen as the basis of dental amalgams? Why does it need to be alloyed with other metals?

   (c) Artificial hearts are made of plastics supported on an aluminium alloy frame. Why is aluminium alloy chosen for this purpose?

   (d) Why was lead chosen for use in plumbing? Why is it no longer used for this job?

7. **X**, **Y** and **Z** are metals. **Y** will displace **X** from the sulphate, **XSO₄**, but will not displace **Z** from **ZSO₄**. Which is the correct order of reactivity of the metals, starting with the most reactive?

   (a) **X, Z, Y**   (b) **X, Y, Z**   (c) **Z, Y, X**   (d) **Y, X, Z**   (e) **Z, X, Y**

8. (a) Why do space ships carry liquid hydrogen and liquid oxygen?

   (b) Why is hydrogen used in meteorological balloons?

   (c) Why is hydrogen called a *clean* fuel?

   (d) What is the reason for converting hydrogen into ammonia?

9. Briefly describe how you would separate copper from a mixture of copper powder and magnesium powder.

10. Nickel and manganese are metals. Describe the experiments you would do to find out which is the more reactive of the two.

Crossword on Chapter 9

Trace or photocopy this grid (teacher, please see note at the front of the book), and then fill in the answers.

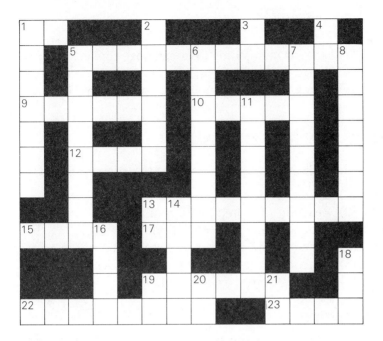

Across

1 Symbol for cobalt (2)
5 This acid reacts with metals (12)
9 A metal which reacts with cold water (6)
10 You can tie these lengths of metal, plastic or animal fibres (5)
12 Hydrogen converts these into fats (4)
13 This low-density metal seems to be unreactive (9)
15 Galvanise with this (4)
17 Timothy (3)
19 Silver things on your bracelet (6)
22 This is what a metal does with electricity (8)
23 A very strong metal (4)

Down

1 A beautiful piece of solid matter (7)
2 Lovely smells (6)
3 Symbol for aluminium (2)
4 Symbol for titanium (2)
5 The gas you get from acids (8)
6 A shiny metal which is used to protect iron (8)
7 Galvanising will help iron in _____ attack by the air (9)
8 A metal which reacts with cold water (7)
11 A very expensive unreactive metal (8)
13 Symbol for astatine (2)
14 This is reflected by polished metals (5)
16 This should tie it up (4)
18 A metal which is used to coat iron (3)
19 County Council (1,1)
20 Symbol for arsenic (2)
21 Symbol for silicon (2)

10 Salts

When hydrogen atoms in an acid are replaced by metal atoms, the substance no longer behaves as an acid: it is a completely new substance with new properties, and it is called a salt.

10.1 Preparation of Salts

Methods which can be used for the preparation of salts are:

(1) Direct combination
(2) Reaction of a metal with a dilute acid
(3) Reaction of a base with a dilute acid
(4) Reaction of a metal carbonate with a dilute acid } Methods for soluble salts
(5) Reaction of an alkali with a dilute acid
(6) Precipitation of an insoluble salt

Direct Combination

In Chapter 4, you saw two elements combine when they were shaken together (aluminium and iodine) or heated together (iron and sulphur). The products, aluminium iodide and iron(II) sulphide, are salts. This method of making salts is called **direct combination**. It is used for reactive elements.

Reaction Between a Metal and an Acid

This method is used to make soluble salts. The reaction is:

Metal + Acid ⟶ Salt + Hydrogen

For example in Experiment 10.1,

Zinc + Sulphuric acid ⟶ Zinc sulphate + Hydrogen

$$Zn(s) + H_2SO_4(aq) \longrightarrow ZnSO_4(aq) + H_2(g)$$

- The method can be used for metals which are above hydrogen in the reactivity series shown in Table 9.2 (p. 154).
- Unreactive metals, e.g., copper and lead, do not react with dilute acids (except dilute nitric acid).
- With very reactive metals, e.g., sodium and potassium, the reaction is too violent to be safe to use.

Reaction Between a Base and a Dilute Acid

This method can be used to make a large number of soluble salts. The reaction is:

Acid + Base ⟶ Salt + Water

For example in Experiment 10.2,

Copper oxide + Sulphuric acid $\longrightarrow$ Copper sulphate + Water

$$CuO(s) + H_2SO_4(aq) \longrightarrow CuSO_4(aq) + H_2O(l)$$

- All metal oxides are bases and react with acids to give salts.
- The method is useful for making the salts of copper and lead and other metals which do not themselves react with acids.
- Sulphuric acid gives sulphates; hydrochloric acid gives chlorides; nitric acid gives nitrates.

Reaction Between a Carbonate and an Acid

This method can be used to make soluble salts. The reaction is:

Metal carbonate + Acid $\longrightarrow$ Metal salt + Carbon dioxide + Water

You used this reaction to make carbon dioxide in Experiment 7.10.

Reaction Between an Alkali and an Acid

The neutralisation of an acid by an alkali can be used to make soluble salts. Alkalis are soluble bases. The reaction is:

Alkali + Acid $\longrightarrow$ Salt + Water

For example, in Experiment 3.5,

Sodium hydroxide + Hydrochloric acid $\longrightarrow$ Sodium chloride + Water

$$NaOH(aq) + HCl(aq) \longrightarrow NaCl(aq) + H_2O(l)$$

- The common alkalis are sodium hydroxide, potassium hydroxide, calcium hydroxide and ammonia solution. (Experiment 10.3 shows how to make an ammonium salt, ammonium sulphate, which is a valuable fertiliser.)
- The neutralisation method is useful for making the salts of sodium and potassium. These metals react so violently with acids that you cannot use the Metal + Acid method to make their salts.

Precipitation

The precipitation method is used to make insoluble salts. Suppose you want to make the insoluble salt barium sulphate. These are the steps to take.

(1) Make a solution of a soluble barium salt, e.g., barium nitrate.

(2) Make a solution of a soluble sulphate, e.g., sodium sulphate.

(3) Mix the two solutions. The insoluble salt barium sulphate is thrown out of solution.

Barium nitrate (solution) + Sodium sulphate (solution) $\longrightarrow$ Barium sulphate (precipitate) + Sodium nitrate (solution)

$$Ba(NO_3)_2(aq) + Na_2SO_4(aq) \longrightarrow BaSO_4(s) + 2NaNO_3(aq)$$

To help you in choosing soluble salts, remember:

- **All nitrates are soluble.**

- **All sodium, potassium and ammonium salts are soluble.**

Experiments 10.4 to 10.6 use the precipitation method to make insoluble salts. Experiment 10.6 tells you how to make some pigments and mix them with what the industry calls a 'vehicle'. This is a mixture of solvent and binder which carries the pigment.

Experiment 10.1

To prepare zinc sulphate

1. Take 50 cm^3 dilute sulphuric acid in a Pyrex beaker. Warm. Add two pieces of granulated zinc.

2. If all the zinc reacts completely and disappears, add more. When the stream of bubbles of hydrogen stops, all the acid has been used up. There will be some zinc left over. You must use an excess of zinc, more than enough to react with all the acid, because it is important to have no acid left at the end of the reaction. Figure 10.1 (a) shows the apparatus.

3. Figure 10.1 (b) shows the next step. Filter to remove the excess of zinc.

4. Evaporate the filtrate until crystals of zinc sulphate just begin to form. Leave to stand. Filter off the crystals of zinc sulphate.

$$\text{Zinc} + \text{Sulphuric acid} \longrightarrow \text{Zinc sulphate} + \text{Hydrogen}$$

$$\text{Zn(s)} + \text{H}_2\text{SO}_4\text{(aq)} \longrightarrow \text{ZnSO}_4\text{(aq)} + \text{H}_2\text{(g)}$$

If you have not made sure that all the dilute sulphuric acid was used up in the reaction, when you start to evaporate the solution of zinc sulphate, the acid will become more concentrated, and there is a possibility of concentrated acid splashing out.

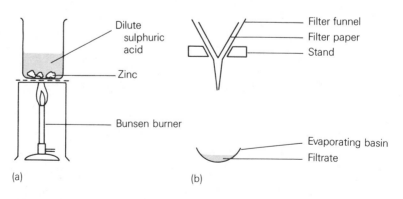

Figure 10.1   Preparation of zinc sulphate

Experiment 10.2

To prepare copper sulphate crystals

1. Use the apparatus shown in Figure 10.1 (a). Into the Pyrex beaker, put 50 cm³ of dilute sulphuric acid. Warm. Add a spatula measure of copper oxide, and stir with a glass rod.

2. If all the copper oxide you have added reacts, add some more. You must have an excess of copper oxide to make sure that all the acid is used up. No gas is evolved in this reaction. You can judge when the reaction is complete because a piece of litmus paper dipped into the solution will no longer turn red when all the acid has been used up.

3. Filter to remove the excess of copper oxide. Put the filtrate of copper sulphate solution into an evaporating dish. Heat until, when you dip in a glass rod and hold it up to cool, small crystals form on the glass rod.

4. Put a few drops of the solution on to a microscope slide. Using either a hand lens or a microscope, watch as crystals form.

5. Leave the rest of the solution to stand. If crystallisation takes place slowly, the crystals will be large.

6. If you obtain some nicely shaped crystals which you want to keep, wrap them up in Sellotape, and stick them into your book or on to a card.

$$\text{Copper oxide} + \text{Sulphuric acid} \longrightarrow \text{Copper sulphate} + \text{Water}$$

$$CuO(s) + H_2SO_4(aq) \longrightarrow CuSO_4(aq) + H_2O(l)$$

Save the rest of the crystals for Experiment 10.8.

Experiment 10.3

To prepare ammonium sulphate

1. Take 50 cm³ dilute sulphuric acid in a beaker, as shown in Figure 10.2 (a).

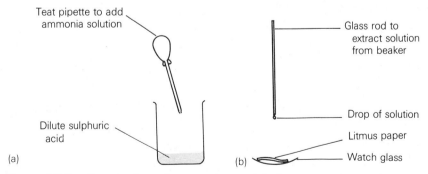

Figure 10.2   Preparation of ammonium sulphate

2. With a teat pipette, add dilute ammonia solution. Stir with a glass rod, and then take out a drop of solution and drop it on to a piece of litmus paper, as shown in Figure 10.2 (b). The paper will turn red, showing that the solution is still acid.

3. Add some more dilute ammonia solution, stir, take out a drop of solution and test it again. Continue adding more ammonia solution until the litmus turns blue, showing that the solution is alkaline. All the acid has been neutralised, and there is an excess of ammonia present. This does not matter as it will be driven off as ammonia gas and steam when you begin to heat the solution.

4. Put the solution of ammonium sulphate into an evaporating basin and heat. When the solution appears ready to crystallise, leave it to stand. Collect the crystals of ammonium sulphate formed.

$$\text{Ammonia} + \text{Sulphuric acid} \longrightarrow \text{Ammonium sulphate}$$

$$2NH_3(aq) + H_2SO_4(aq) \longrightarrow (NH_4)_2SO_4(aq)$$

Experiment 10.4

To prepare some pigments and make them into coloured paints

1. With a measuring cylinder, measure into separate beakers the stated volumes of the two solutions required for the pigment.

2. Mix the solutions by stirring.

3. Fit a clean Buchner funnel into a filter flask, as shown in Figure 10.3, and attach a filter pump to the flask. Lay a filter paper in the funnel, wet it, and apply gentle suction. Pour the contents of the beaker into the funnel, increasing suction if necessary to draw all the liquid through. Wash the precipitate with water. (If you do not have a Buchner funnel, you can use a filter funnel and filter paper; it just takes longer.)

4. When all the water has passed through, spread the damp pigment on a paper towel to dry.

5. The volumes of solutions (all 10% concentration) to mix are:
   (a) For white pigment, basic lead carbonate, add 75 cm^3 lead nitrate solution to 15 cm^3 sodium carbonate solution.
   (b) For yellow pigment, lead chromate, add 15 cm^3 potassium chromate solution to 75 cm^3 lead nitrate solution.
   (c) For Prussian blue, iron(III) hexacyanoferrate(II), add 20 cm^3 iron(III) chloride solution to 75 cm^3 potassium hexacyanoferrate(II) solution.
   (d) For green pigment, basic copper carbonate, add 60 cm^3 copper sulphate solution to 30 cm^3 sodium hydrogencarbonate solution.

6. Take some of the white pigment; break it up with a spatula on a watch glass. In a test tube, mix one part of glue with three parts of water. Add a few drops of this paint 'vehicle' to the pigment and mix with the spatula. add a few more drops and mix well. Repeat until you have a paint of a nice consistency.

Repeat the mixing with the other pigments. Try mixing a coloured pigment with white to make a pastel shade. When you have assembled a few coloured paints, take a paint brush and paint a picture with them.

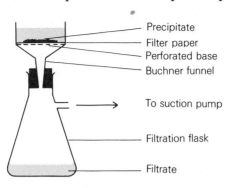

Figure 10.3   Filtration of precipitates

Experiment 10.5

Making a salt by precipitation

1. Make a solution of tannic acid. Boil 1 teabag in $50\ cm^3$ of water. Let the solution cool. Pour off the tea. This solution contains tannic acid.

2. Dissolve a crystal of iron(II) sulphate in $10\ cm^3$ of water. Add this solution to the tea. Filter.

3. Describe the precipitate you obtain. Copy and complete the word equation,

Tannic acid + Iron(II) sulphate $\longrightarrow$ _____ _____ + _____ _____

Experiment 10.6

A use for precipitation

1. Dissolve about 5 g of sodium phosphate (Na_3PO_4) in water.

2. Add limewater (calcium hydroxide solution).

3. Can you name the precipitate?
   Copy and complete the equation:

   Sodium phosphate + Calcium hydroxide $\longrightarrow$ _____ + _____

   Say which of the two products is the insoluble one.

4. This reaction is used to rid sewage of phosphates. Would a similar method work for nitrates? Explain your answer. Why do people want to remove phosphates and nitrates from treated sewage?

Experiment 10.7

Growing silicates

Most silicates are insoluble. Sodium silicate, however, is a soluble compound which is sold as *waterglass*.

1. Make a solution of 50 cm^3 of waterglass in 250 cm^3 of water. Pour the solution into a jam jar. (Beakers are difficult to clean after this experiment.)

2. Cover the bottom of the jar with a thin layer of sand.

3. Drop in crystals of iron(II) sulphate, cobalt chloride, nickel sulphate, copper sulphate, manganese sulphate, chromium sulphate and any other coloured soluble salts you can find.

4. Leave the crystals overnight. Can you see what people describe as a 'chemical garden'?

   One of the reactions that have happened is:

$$\begin{array}{ccc} \text{Copper} & \text{Sodium} & \text{Copper} & \text{Sodium} \\ \text{sulphate} + \text{silicate} \longrightarrow \text{silicate} & + \text{sulphate} \\ \text{(solution)} & \text{(solution)} & \text{(insoluble)} & \text{(solution)} \end{array}$$

   Can you explain the reactions of (a) nickel sulphate and (b) chromium chloride?

10.2 Crystals

Water of Crystallisation

Many salts crystallise as **hydrates**, crystals which contain water. Hydrates always have definite formulas. In magnesium sulphate crystals, there are always 7H$_2$O for every MgSO$_4$, and the hydrate therefore has the formula MgSO$_4$.7H$_2$O. It is called magnesium sulphate-7-water, but is better known as Epsom salts. Copper sulphate crystals have the formula CuSO$_4$.5H$_2$O, and are named copper sulphate-5-water. The combined water gives both colour and shape to the crystals. It is called **water of crystallisation**.

Growing Large Crystals

The more slowly crystals form, the larger they become. Methods of obtaining large crystals are:

- Filter the solution to get rid of dust. Otherwise many small crystals of solute will form around dust particles.

- Lag the beaker holding the solution to slow down the rate of cooling.

- Slow down the rate of evaporation by covering the solution with a piece of paper pierced with a few holes.

- Hang in the solution a small 'seed' crystal for the solute to crystallise around.

Experiment 10.8

Heating copper sulphate crystals

1. Take some copper sulphate crystals and put them into the side-arm boiling tube shown in Figure 10.4.

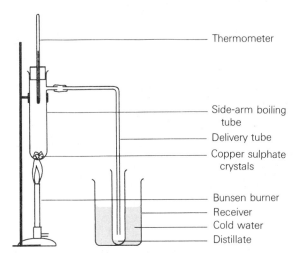

Figure 10.4   Heating copper sulphate crystals

2. Warm with a small flame. Watch the thermometer. If you heat strongly, and the temperature soars suddenly, the sudden expansion of mercury may break the thermometer.

3. Collect any distillate in the receiver. What is the thermometer reading as the liquid distils over? Which liquid has this boiling point?

4. What changes do you see in the colour and shape of the crystals after heating? What do you think has caused the changes?

5. The substance left in the side-arm tube is called **anhydrous copper sulphate**. When the anhydrous copper sulphate is cool, add a few drops of water to it. You will notice *two* changes. Make a note of them.

6. Have you worked out what *anhydrous* means? If you have, you can complete the equation:

    Copper sulphate-5-water ⟶ _____ + _____

    $CuSO_4.5H_2O(s)$ ⟶ _____ + _____

   What do your observations in step (5) tell you about this reaction?

7. What do you think will happen if you heat crystals of cobalt chloride-6-water, which are pink? Try it and see whether you are right!

Experiment 10.9

To grow a large crystal of alum

1. Alum is the common name for aluminium potassium sulphate. This forms crystals with water of crystallisation, producing $KAl(SO_4)_2.12H_2O$.

2. Prepare a solution of alum. Take a 250 cm^3 beaker and put into it 150 cm^3 water. Warm, but do not boil. Add alum and stir. When no more alum will dissolve, allow the solution to cool. You now have a cold, saturated solution.

3. Filter the solution into a clean beaker. Take 5 drops of the solution on a microscope slide. Examine the solution either with a hand lens or under a microscope. You will see crystals growing.

4. Take a small, well shaped crystal of alum. Suspend it by a thread from a glass rod so that the crystal hangs in the middle of the solution (see Figure 10.5).

5. Cover with a sheet of paper pierced with a few holes. Set aside in a place where the temperature is constant.

6. The solution will slowly evaporate, and the small crystal will grow in size. If small crystals form elsewhere, remove them. As the level of the saturated solution falls, top it up with more. If the solution is not saturated, the crystal will dissolve.

7. There are various ways of preserving the crystal you have grown, to prevent it losing its water of crystallisation. After drying it, you can put the crystal into a specimen bottle with a plastic top. You can coat the crystal with perspex cement or nail varnish and mount it on a card, or you can seal it into a glass tube.

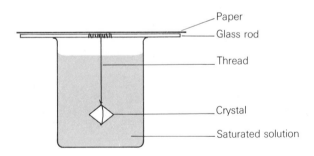

Figure 10.5   Growing a crystal

10.3 Some Useful Salts

Common Salt

The salt we call 'common salt' is sodium chloride. It has been mined for thousands of years. Before the days of refrigeration, it filled a vital need as a food preservative. It takes moisture out of meat and fish.

The dried food can be kept much longer before it spoils. For centuries, farmers used to kill cattle and sheep before the onset of winter. People ate salted meat until the following summer.

We need to eat salt. Salt takes part in muscle action, in the carrying of messages along nerve fibres and in digestion. Without salt, paralysis sets in, and death follows. Through perspiration, the body loses both water and salt. We have built-in controls to regulate our salt content. If we eat too much salt, our kidneys excrete it; if we eat too little, our kidneys excrete water but no salt.

Emperor Napoleon I of France led his troops into Russia in 1812. The harsh Russian winter was something he was not prepared for, and he was forced to retreat. Salt starvation was one of the hardships his soldiers had to endure. It lowered their resistance to disease, and epidemics spread. Wounds which might otherwise have healed stayed open and became infected. Thousands died.

Salt is an important raw material in many branches of the chemical industry. From salt are obtained:

> chlorine – used for disinfecting water and in the manufacture of plastics, e.g. PVC
>
> sodium – used as a coolant in nuclear reactors
>
> hydrogen – used in the manufacture of fertilisers and margarine
>
> sodium carbonate – used in the manufacture of glass and as washing soda
>
> sodium hydrogencarbonate – used in indigestion tablets and as baking soda
>
> mild antiseptics, e.g. Milton®
>
> powerful weedkillers, e.g. Tandol®

Experiment 10.10

Why is salt spread on the roads in winter?

When the roads are covered with ice, you see trucks spreading salt. What effect does salt have on ice? Can you think of an experiment to find out?

1. You are supplied with two large filter funnels with stands, crushed ice, two thermometers, some coarse commercial salt, two large beakers, two measuring cylinders, glass rods and any other common laboratory apparatus you need.

2. Select the apparatus you need. Sketch the set-up you plan to use. Briefly say what you plan to do. When your teacher has approved your plan, carry out your experiment.

3. Say what property of ice is affected by salt. Explain how this action of salt on ice leads to the use of salt on icy roads.

Silver Bromide

A black-and-white photographic film is a piece of celluloid coated with a film of gelatin containing silver bromide. When light falls on silver bromide, the salt becomes *sensitised* to reducing agents, that is, more easily converted into silver. Figure 10.6 shows what happens when a film is exposed to light.

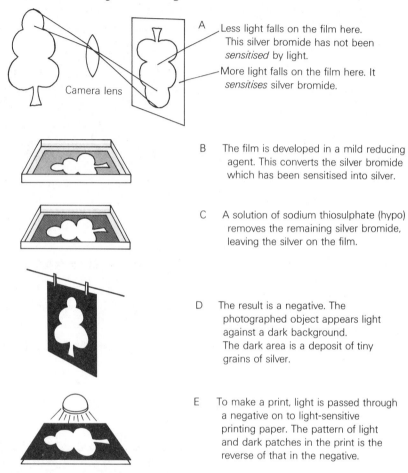

A — Less light falls on the film here. This silver bromide has not been *sensitised* by light.

More light falls on the film here. It *sensitises* silver bromide.

Camera lens

B — The film is developed in a mild reducing agent. This converts the silver bromide which has been sensitised into silver.

C — A solution of sodium thiosulphate (hypo) removes the remaining silver bromide, leaving the silver on the film.

D — The result is a negative. The photographed object appears light against a dark background. The dark area is a deposit of tiny grains of silver.

E — To make a print, light is passed through a negative on to light-sensitive printing paper. The pattern of light and dark patches in the print is the reverse of that in the negative.

Figure 10.6   Black-and-white photography

Phosphate Fertilisers

In the eighteenth century, farmers were in the habit of using crushed bones as a fertiliser. The bone manure worked slowly. It had to remain on the ground until slow natural processes turned the chemicals in bone into soluble compounds which could reach the plants. A British farmer called John Bennett Lawes began to study bone manure in 1840. In Germany, Justus von Liebig (of condenser fame) was working along the same lines. Both men knew that the valuable ingredient in bone dust was calcium phosphate. They were after a way of making it easier for the plants to absorb it. In 1840, Lawes and Liebig discovered that calcium phosphate reacted with sulphuric acid to form a soluble salt. Plants could feed on this soluble fertiliser as soon as it was applied. Lawes was the first to patent the discovery. He called the new fertiliser 'superphosphate of lime'.

A NEW FIRST CHEMISTRY COURSE

British farmers took to the new product with enthusiasm. It increased their crops fivefold. Soon, fertiliser manufacturers were using more bones than the butchers could supply. By the 1850s, Britain began importing phosphate-containing rock from South America. The ore 'rock phosphate' is still the source of phosphate fertilisers. Now, it is converted into ammonium phosphate. Besides being very soluble, this salt has the advantage of supplying the crops with nitrogen as well as phosphorus. (Chapter 5 dealt with the importance of nitrogen compounds.)

Potassium Chloride

The fertilisers called NPK fertilisers contain nitrogen compounds (nitrates and ammonium salts), phosphorus compounds (phosphates) and potassium salts (usually potassium chloride). You read about the manufacture of nitrates and ammonium salts in Chapter 5. You have just read about the manufacture of phosphates. Potassium chloride does not need to be manufactured as it is mined near Whitby in Yorkshire.

Calcium Fluoride

Fluorine is a poisonous yellow gas. It is a non-metallic element. It forms salts called *fluorides*. Calcium fluoride is a compound of calcium and fluorine.

Tooth decay is a serious problem. Dentists noticed that children in regions where the water contains fluorides have fewer cavities than children in regions where the natural fluoride content of water is low. This is why people have started using toothpastes which contain calcium fluoride. In some parts of the country, the water treatment works add fluoride to the drinking water. This is to ensure that everyone gets protection from dental decay. The people who benefit most are children up to the age of eight. Their teeth are still calcifying. There is a danger, however. The concentration of fluoride in drinking water should not be more than one part per million (one ten-thousandth of a percent). If too much fluoride is present, it *causes* tooth decay. For this reason, some people worry over the fluoridation of the water supply. In 1985, the UK Government authorised the Water Boards to add fluoride to drinking water.

In Grafton, Australia, a pro-fluoride float followed an anti-fluoride float in the annual flower festival procession. A brawl broke out, and one man was whipped and another stabbed. A politician who supported fluoridation of the drinking water received a threatening letter and a bomb through the post. These incidents show how high feelings run on the subject of fluoridation. In fact, 80 per cent of the Australian population drinks fluoridated water.

Plaster of Paris

Calcium sulphate is mined as *gypsum*, $CaSO_4.2H_2O$. When it is heated to 110 °C, it forms a compound of formula $CaSO_4.\frac{1}{2}H_2O$, which is called *plaster of Paris*. When mixed with water, this sets to form a hard mass of the hydrate $CaSO_4.2H_2O$, and expands slightly. This is why plaster of Paris is used for making plaster casts for broken limbs. It is also used for plastering walls.

Experiment 10.11

Plaster of Paris

1. Mix plaster of Paris with enough water to make a paste like medium thick cream. Pour the paste into a small box (e.g. matchbox).

2. Smear a coin or other object lightly with grease. Press it into the plaster, and leave it for 2–3 hours. Then lift the coin out, as in Figure 10.7.

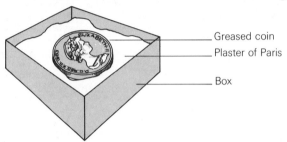

Figure 10.7   Making a plaster cast

3. How long does it take for plaster of Paris to set hard?

   (a) Weigh 50 g of plaster of Paris in a paper cup.

   (b) Take 30 cm³ of water in a measuring cylinder.

   (c) Pour the water into the plaster of Paris, stir, and start a stop clock.

   (d) Pour the mixture into a disposable plastic Petri dish, and level it off with a ruler.

   (e) When the plaster seems to be setting, test it with a weight (about 40 g). This will dent soft plaster but will not sink into the plaster after it has hardened. Note the time taken for the plaster to set hard.

   (f) Repeat the test with the volumes of water shown in the table. Copy the table, and fill in your results.

Volume of water (cm³) added to 50 g of plaster of Paris	Time to set hard
30	
40	
50	
60	

   (g) Plot your results on graph paper.

   (h) How does the volume of water used affect the strength of the plaster? It is up to you to design an experiment to test the strength of the samples of plaster that you have made.

   (i) What effect does it have on the setting time if you add 3 g of alum (aluminium potassium sulphate) to the plaster?

Sodium Carbonate and Sodium Hydrogencarbonate

Sodium carbonate crystallises from solution as a hydrate, $Na_2CO_3.10H_2O$. Solutions of sodium carbonate are strongly alkaline. This helps them to remove grease and dirt from fabrics, and gives sodium carbonate the name of **washing soda**. Washing powders contain sodium carbonate and perfume and colouring material.

Sodium hydrogencarbonate solutions are weakly alkaline. Being safe to swallow, sodium hydrogencarbonate is used in indigestion powders to neutralise stomach acidity. It is referred to as *sodium bicarbonate* or 'bicarb of soda'.

Sodium hydrogencarbonate decomposes readily on gentle heating to give carbon dioxide. This is why it is used as a rising agent in bread and cake mixtures. It is known as *baking soda*. Baking powder contains sodium hydrogencarbonate and tartaric acid, a weak acid. The equation for the decomposition of sodium hydrogencarbonate is

$$\text{Sodium hydrogencarbonate} \longrightarrow \text{Sodium carbonate} + \text{Carbon dioxide} + \text{Steam}$$

$$2NaHCO_3(s) \longrightarrow Na_2CO_3(s) + CO_2(g) + H_2O(g)$$

Sodium carbonate does not decompose on heating.

Experiment 10.12

To make some bath salts

1. Take a beaker (250 cm³) containing 100 cm³ water. Warm it on a tripod and gauze.

2. Add sodium carbonate, stirring with a glass rod, until no more will dissolve.

3. Filter. Add a little colouring matter (for example cochineal). Leave the solution in a crystallising dish covered with a piece of paper pierced with some holes.

4. After a few days, you will have a batch of crystals. Filter, and spread them on a paper towel to dry.

5. Put the rest of the crystals into a bottle. Add a little perfume, and shake very gently to distribute the perfume without smashing the crystals.

6. Stopper the bottle of bath salts.

Experiment 10.13

How fast do Alka-Selzer tablets react with water?

People take Alka-Selzer tablets for indigestion. The tablets react with water to give a gas.

1. What is the gas? Can you devise a method of collecting the gas and testing it? One quarter of a tablet will give you enough gas to test.

   When your teacher has approved your plan, test the gas.

2. How fast do Alka-Selzer tablets react with water? Sketch an apparatus in which you could (a) allow a half-tablet of Alka-Selzer to react with water and (b) collect and measure the gas formed. When you are satisfied with your design, show it to your teacher. If he or she approves, use your apparatus for the following measurements.

3. Drop a half-tablet into water. Measure the volume of gas collected after 10 seconds, 20s, 30s, 40, 60s and 80s. Tabulate your results. On graph paper, plot the volume of gas collected against the time interval.

Time (seconds)	Volume of gas (cm³)
10	
20	
30	
40	
60	
80	

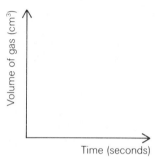

What do you notice about the shape of your graph?

4. Find out how fast gas is formed when you break up a half-tablet into small pieces before dropping it into water.

5. Grind a half-tablet into powder and repeat the measurement.

6. What happens if you warm the water first?

Anhydrous sodium carbonate, Na_2CO_3, is used in large quantities by the glass industry. The first glass was made by melting sand, which is silicon(IV) oxide, SiO_2. When molten sand was allowed to cool, a transparent solid was formed. Sand has a high melting point (1710 °C). Thousands of years ago, the ancient Egyptians found out that it would melt more easily if they added other minerals to it. When they melted together sand and calcium carbonate and sodium carbonate, they obtained glass. Egypt is one of the few places where sodium carbonate occurs naturally.

The same materials, sand and calcium carbonate and sodium carbonate, are still used for making glass. For special purposes, other ingredients are added.

- For coloured glass, a metal oxide is added, e.g., cobalt oxide for blue glass, chromium oxide for green glass, and manganese oxide for purple glass.
- The addition of lead oxide gives a glass with a beautiful shine.
- Glass can be made stronger by adding boron oxide. This enables glass to stand up to heat as well as to blows. Pyrex glass is made in this way.

- Light-sensitive glass, which darkens in bright light and returns to its original state in dimmer light, is made by adding silver choride.

- Glass ceramic, which is almost unbreakable, is made by heating light-sensitive glass for some hours. It is used in rocket nose-cones and space-shuttle tiles.

Questions on Salts

1. (a) What part did salt play in Columbus' voyage of discovery?

   (b) Why is salt especially valued in hot climates?

   (c) What effect did a shortage of salt have on Napoleon's troops?

2. What is the connection between salt and (a) PVC raincoats   (b) a disinfectant used for sterilising babies' bottles   (c) the weeds on the garden path   (d) washing soda?

3. What raw materials are needed for the manufacture of glass? Why were the Egyptians the first people to make it?

4. Which sort of glass would you choose for the manufacture of (a) ovenware (b) sunglasses (c) laboratory glassware (d) stained glass windows   (e) a sparkling flower vase   (f) a space shuttle (g) glass windows?

5. What are the chemical differences between washing soda and baking soda? How do the chemical differences lead to different uses for the salt?

6. What is the difference between a photograph and a photographic negative? How can you make a print from a negative?

7. Give three uses for plaster of Paris. Explain why the salt can be used for the purposes you have mentioned.

8. What effect does calcium fluoride have on the human body?

9. Why is the manufacture of phosphate fertiliser so important? What did farmers use before there was a fertiliser industry? What was the disadvantage of the natural fertiliser they used? How did the manufactured fertiliser improve on this? Which phosphate is used nowadays? What advantages does it have over previous fertilisers?

Questions on Chapter 10

1. Say whether each of the following is an acid or a base or a salt:

   (a) zinc oxide   (b) limewater
   (c) ammonia   (d) carbon dioxide solution
   (e) copper sulphate   (f) lead nitrate
   (g) magnesium oxide   (h) sulphur dioxide solution

2. What do all acids contain? What is a salt? Name the salts that are made from (a) hydrochloric acid   (b) sulphuric acid and   (c) nitric acid.

3. Where is salt mined in the UK? How is it mined? (See Chapter 2 if you need help.)

4. Describe how you would prepare:

   (a) a salt of zinc, starting from the metal

   (b) a salt of sodium, starting from sodium hydroxide.

5. Write down the correct words to fill the gaps:

$$\text{Acid} + \text{Alkali} \longrightarrow \text{Salt} + \underline{\hspace{2cm}}$$

$$\text{Acid} + \text{Base} \longrightarrow \text{Salt} + \underline{\hspace{2cm}}$$

$$\text{Acid} + \text{Metal} \longrightarrow \text{Salt} + \underline{\hspace{2cm}}$$

$$\text{Acid} + \text{Carbonate} \longrightarrow \text{Salt} + \underline{\hspace{2cm}} + \underline{\hspace{2cm}}$$

6. (a) Name two solutions which you could mix to make a precipitate of the insoluble salt silver bromide.

   (b) What experiment could you do to show that the silver bromide you have made turns black in sunlight but not in the absence of sunlight?

   (c) What use is made of the effect of light on silver bromide?

7. Blue copper sulphate crystals are heated.

   (a) What is the final colour after heating?

   (b) What liquid can be collected from the crystals? How can you identify it?

   (c) What name is given to this liquid when it is part of the crystals?

   (d) What is the name of copper sulphate without the liquid?

   (e) Is the change that happens reversible or irreversible? Explain your answer.

8. (a) What compound is present in plaster of Paris? What happens when you add water to it?

   (b) What use do we make of calcium fluoride?

   (c) What is the chemical name for bath salts?

   (d) What is sodium hydrogencarbonate used for?

   (e) How can you tell the difference between sodium carbonate and sodium hydrogencarbonate?

9. What are the three types of ingredient in NPK fertilisers? Explain how they are obtained or made.

Crossword on Chapter 10

Trace or photocopy this grid (teacher, please see note at the front of the book), and then fill in the answers.

Across

1 See 17 down
5, 14 down One starting material in the preparation of a salt (5,5)
8 The highest point (3)
10 Roman house (5)
12, 3 down, 13 down They turn from blue to white on heating (6,8,8)
14 '_____ aye!' says the Scotsman (3)
16 See 4 down
17 Symbol for beryllium (2)
21 One way of making salts (14)
23 See 4 down
24 Some of these salts are fertilisers (8)

Down

2 Ending to a morning's work (5)
3 See 12 across and 11 down
4, 23 across, 16 across Lost when 12 across, 3 down, 13 down is heated (5,2,15)
6 A starting material in the preparation of salts (4)
7, 20 down A name for sodium carbonate-10-water (7,4)
9 Alternatively (2)
11, 3 down Plaster of Paris (7,8)
13 See 12 across
14 See 5 across
15 Colour of anhydrous cobalt chloride (4)
17, 1 across A use for sodium carbonate crystals (4,5)
18 Salts must be heated very strongly before they do this (4)
19 A laboratory should be this (4)
20 See 7 down
22 Decay (3)

Index